Culturally Responsive Science Pedagogy in Asia

Science learning, for many, is often seen as learning a culture of science knowledge and practices that is incongruent from one's everyday experiences and cultural background of learners. This edited volume presents a systemic view of the current initiatives and challenges for the inclusion of culturally responsive science pedagogy (CRSP) in non-Western and multicultural contexts in three Asian countries – Malaysia, Indonesia and Japan.

Split into three parts, the book examines the history and current educational systems, curriculums and sociocultural diversities in each country, offering an updated review of equity in education. It reflects and expands on the role of CRSP in diverse societies before going into case studies that feature the experiences of teachers in implementing CRSP in Malaysia, Indonesia and Japan. These snapshots reflect the multiple ways equity is addressed in the teaching and learning of science in these Asian countries, allowing readers to extrapolate the possible challenges and best practices for designing and implementing CRSP in practice. The final section examines how these findings provide a sustainable platform for building capacity in understanding the cultural complexities and realities of recruiting and retaining diverse students into science.

One of few books to investigate the role of CRSP in diverse societies in Malaysia, Indonesia and Japan, this book makes a unique contribution to the field of science education with reference to culturally responsive pedagogy. Its strategies and solutions serve as an important comprehensive reference for researchers and science teacher educators.

Lilia Halim is Professor of Science Education in the Center of STEM Enculturation, Faculty of Education, Universiti Kebangsaan Malaysia. Her research interests include the development of teachers' professional knowledge and STEM and science education in formal and informal contexts.

Murni Ramli is Senior Lecturer in Educational Sciences and Curriculum Management at the Department of Biology Education, Faculty of Teacher Training and Education, Universitas Sebelas Maret (UNS), Indonesia. She serves as the coordinator for international students' services at the International Office of UNS.

Mohd Norawi Ali is Senior Lecturer in Science Education at the School of Educational Studies (SES), Universiti Sains Malaysia (USM). His research interests are project-based STEM education and ICT integration in science teaching for pre-service teachers and social wellbeing of family and community. He was the editor for *Digest Pendidik Journal*, SES, USM.

Routledge Series on Schools and Schooling in Asia

For the full list of titles in the series, please visit: www.routledge.com/Routledge-Series-on-Schools-and-Schooling-in-Asia/book-series/RSSSA

Culturally Responsive Science Pedagogy in Asia

Status and Challenges for Malaysia, Indonesia and Japan

Edited by Lilia Halim, Murni Ramli and Mohd Norawi Ali

Routledge
Taylor & Francis Group

LONDON AND NEW YORK

First published 2023
by Routledge
4 Park Square, Milton Park, Abingdon, Oxon OX14 4RN

and by Routledge
605 Third Avenue, New York, NY 10158

Routledge is an imprint of the Taylor & Francis Group, an informa business

© 2023 selection and editorial matter, Lilia Halim, Murni Ramli, and Mohd Norawi Ali; individual chapters, the contributors

The right of Lilia Halim, Murni Ramli, and Mohd Norawi Ali to be identified as the authors of the editorial material, and of the authors for their individual chapters, has been asserted in accordance with sections 77 and 78 of the Copyright, Designs and Patents Act 1988.

British Library Cataloguing-in-Publication Data
A catalogue record for this book is available from the British Library

Library of Congress Cataloging-in-Publication Data
A catalog record for this book has been requested

ISBN: 978-0-367-76768-6 (hbk)
ISBN: 978-0-367-76826-3 (pbk)
ISBN: 978-1-003-16870-6 (ebk)

DOI: 10.4324/9781003168706

Typeset in Galliard
by Apex CoVantage, LLC

Contents

Illustrations

Figures

Tables

Contributors

Nurazidawati Mohamad Arsad is Senior Lecturer at STEM Enculturation Centre, Faculty of Education, Universiti Kebangsaan Malaysia (UKM). Her research focus is on STEM education and culturally responsive pedagogy towards marginalised groups.

Muhammad Abd Hadi Bunyamin is Senior Lecturer in Physics Education at Universiti Teknologi Malaysia (UTM). His research interests are in funds of knowledge in physics and integrated STEM education.

Hartini Hashim is a senior science teacher at Sek Men Sultan Ismail, Kelantan. Currently, she is Deputy President of National STEM Association (NSA) Chapter Kelantan. Her research interest is in nurturing entrepreneur thinking in STEM project.

Izzah Mardhiya Mohammad Isa is a doctoral candidate of physics education at UTM. Her research interests include social justice, equity and cultural studies in physics education.

Siti Nur Diyana Mahmud is a senior lecturer at the UKM, and her research interests are in teaching and learning innovation in science and environmental education, systems thinking and science teacher education.

Nurfaradilla Mohamad Nasri is an expert in curriculum and pedagogy. Her main research interest relates to the development of culturally responsive curriculum and instruction, teachers' professional development, and self-directed learning.

Kiyoyuki Ohshika is Professor of Science Education at Aichi University of Education, Japan. His research interests are in the area of curriculum development and development of teaching materials in science for educational institutions such as museums and zoo.

Mohd Ali Samsudin is Associate Professor of Digital STEM education at Universiti Sains Malaysia (USM). His research interests are artificial intelligence in education, STEM digital learning, social network analysis and computerized adaptive testing.

Edy Hafizan Mohd Shahali is Senior Lecturer at the Faculty of Education, Universiti Malaya. His research interests are in integrated STEM education and teachers professional development.

Bhaskar Upadhyay is Associate Professor of Science Education in the Department of Curriculum and Instruction at the University of Minnesota, Twin Cities. His research focuses on equity and social justice in STEM and science education.

Series Editor Note

The so-called Asian century is providing opportunities and challenges for the people of both Asia and the West. The success of many of Asia's young people in schooling often leads educators in the West to try and emulate Asian school practices. Yet these practices are culturally embedded. One of the key issues to be taken on by this series, therefore, is to provide Western policymakers and academics with insights into these culturally embedded practices in order to assist in better understanding of them outside of specific cultural contexts.

There is vast diversity as well as disparities within Asia. This is a fundamental issue and for that reason, it will be addressed in this series by making these diversities and disparities the subject of investigation. The 'tiger' economies initially grabbed most of the media attention on Asian development, and more recently China has become the centre of attention. Yet there are also very poor countries in the region, and their education systems seem unable to be transformed to meet new challenges. Thus, the whole of Asia will be seen as important for this series in order to address questions relevant not only to developed countries but also to developing countries. In other words, the series will take a 'whole of Asia' approach.

Asia can no longer be considered in isolation. It is as subject to the forces of globalisation, migration and transnational movements as are other regions of the world. Yet the diversity of cultures, religions and social practices in Asia means that responses to these forces are not predictable. This series, therefore, is interested to identify the ways tradition and modernity interact to produce distinctive contexts for schools and schooling in an area of the world that impacts countries across the globe.

Against this background, the current volume, dealing with culturally relevant approaches to science pedagogy, makes a welcome addition to the Routledge Series on Schools and Schooling in Asia. It explores the ways science education can be made more relevant for students when account is taken of culturally meaningful processes and content that can support student learning and achievement.

Kerry J. Kennedy
Series Editor
Routledge Series on Schools and Schooling in Asia

Preface

This book argues for culturally responsive science pedagogy (CRSP) to be the framework in the enactment of the science curriculum, pedagogy and assessment for Asian countries. The call for CRSP has never been important than before for two reasons: (a) the decline in number of students participating in science at all levels of education, regardless of students' cognitive abilities, interests and backgrounds – this phenomenon is experienced by countries all over the world; and (b) the need to be more proactive in terms of actions in solving science and technological problems for the benefits of the environment and society. Based on these two premises, this edited book has three parts: (a) Concepts and Contexts, (b) Case Studies and (c) Way Forward.

Chapters 1–5 under Part I "Concepts and Contexts" first aim to set the theoretical understanding of relevant theories underlying CRSP. Bhaskar Upadhyay in Chapter 1 argues:

> An essential feature of CRSP is social action and the desire for social change. Any science teaching and learning engagement must focus on the premise that the learners will gain skills and knowledge to ask questions and generate problems that critically reflect and examines social and cultural discrimination.

In other words, science learning is not for academic purposes only. CRSP is often discussed in the Western literature as a tool to engage and understand students from different cultures (i.e. immigrants) to the major culture of the state. Very few research and writings investigate the role of CRSP in diverse societies such as Malaysia and Indonesia. The geography of each country (urban, rural and interior) further creates subcultures within the existing major culture of a country. Also, teaching science is often from the Western orientation, and, thus, in practice the influence of cultural background is not addressed widely and explicitly. Consequently, understanding how CRSP is relevant in Malaysia, Indonesia and Japan is explored in Chapter 2. Nurfaradilla Mohamad Nasri then addresses the interrelation of equity, equitable and CRSP for the context of Malaysia, Indonesia and Japan in Chapter 3. Chapters 4 and 5 focus on the interlink between relevant educational theories and models with CRSP. In particular, in Chapter 4, Edy,

Lilia and Mohd Ali argue that the pedagogical-related constructs – namely, students' prior knowledge, funds of knowledge and contextual teaching and learning – should be intentionally linked to the concepts of CRSP. Chapter 5, by Mohd Norawi and Hartini Hashim, explicitly showcased how funds of knowledge are associated with CRSP. In addition, a plethora of funds of knowledge from the Asian contexts – namely, Malaysia and Indonesia – are provided in the chapter as references for science teachers. Chapter 5 also explicitly operationalised the concept of culture identified in Chapter 1: culture is non-static and it is a social construct – thus students' culture evolves and develops with their everyday interactions with the community, peers and others.

Part II of the book relates to "Case Studies" in employing CRSP in Indonesia, Malaysia and Japan, respectively. CRSP has shown promise, but many teachers have struggled in conceptualising CRSP and how to implement it in the science classroom. In the Indonesian context (Chapter 6), Murni Ramli acknowledges that Indonesia has begun to consider CRSP in their science education policy as described in the 2013 curriculum. Nevertheless, Murni argues that research on CRSP in Indonesia should be intensified and the research links and matches between schools and universities need to be sharpened and strengthened so that the training for teachers to promote CRSP is informed by research and practice towards improving the quality of CRSP-based learning in Indonesia. In the case of Malaysia (Chapter 7), students found that CRSP enables them to understand science better and more relevant to their prior knowledge. But Siti Nur Diyana Mahmud argues that teachers not only need to have but also need to adopt an innovative mindset for CRSP to be successfully implemented in the classroom in a sustainable way. For the case of Japan, Kiyoyuki Ohshika and Murni (Chapter 8) have shown the importance of science learning towards social action and change at the local, community and national levels. This component of CRSP is normally a consequence of academic success. However, the Japanese education system sees it as an integrated component of the science learning that aims to bring relevance not only as early as possible to students but also as proactive agents to providing solutions to problems that might be due to social and cultural discrimination. Examining the problems critically and solving complex problems imbued by the economy, social and political systems lead to students acquiring the twenty-first-century skills – communication, collaboration, inquiry skills, creativity and complex problem-solving skills, including related disposition for realising equity and social justice in the system.

Part II of this edited book – "Way Forward" – not only summarises the concept of CRSP and how to support CRSP in a sustainable way (Chapter 10) but also provides a way forward for STEM education with an associated CRSP-related variable, that is, gender (Chapter 9). Muhammad Abd Hadi and Izzah Mardhiya argue that there is a need to address gender equity issues in real school contexts, balancing perspectives on racial equity with gender equity when doing research that connects the two and promoting parents as critical agents that can influence gender equity issues at the community level. Nurazidawati, Nurfaradilla and Siti Nur Diyana in Chapter 10 have highlighted the need for teachers to ensure constructive alignment between teaching, learning and assessment and the underlying value for this constructive alignment is CRSP. CRSP values should also underpin the preparation and professional development of science teachers.

As a conclusion, in this edited volume, the editors combine the authentic voices of authors from different contexts and cultural worldviews to assimilate and synthesise broad theoretical concepts of CRSP and provide guidance on how to transform CRSP as a pedagogical framework in practice. The purpose of this edited volume is to provide educators and graduate students/scholars in the field of education with the knowledge, skills and dispositions to facilitate success for *all* students. Comparative studies between Malaysia, Indonesia and Japan can inform researchers of what to imitate and avoid in future planning for policy and curriculum development and implementation regarding CRSP. As summed up by Bhaskar Upadhyay, science teachers not only need to be more conscious of but also need to explicitly invite students' culture into science activities and discourses.

Acknowledgements

We would like to convey our utmost appreciation to the Sumitomo Foundation, Japan, for awarding us the grant (Reg Num: 178478; UKM GG-2019–004). The grant has enabled us to create research networking and collaboration among educational researchers from Malaysia, Indonesia, Japan and also the United States. This edited volume is one of the outcomes of the research grant – in view of investigating CRSP practices in each studied country for the betterment of science education as a whole.

Our appreciation also goes to Tuan Mastura Tuan Soh (Malaysia), who has contributed to the earlier stage of the manuscript preparation.

As for data collection, our thanks go to those who have participated in the research work: In Indonesia, the *Himpunan Pendidik dan Peneliti Biologi Indonesia*, HPPBI (Indonesian Society for Biology Educators and Researchers, ISBER), and in Japan, Okazaki Attached Elementary and Junior High School, Chiryu Public Elementary School, Fukuda Elementary School and Jyoto Elementary School.

In Malaysia, our thanks go to Mohd Sharul Anuar bin Omar (*Sek. Keb. Bangsar, Kuala Lumpur*), Nadirah binti Kamarudzaman (*SK Putrajaya Presint 16, Putrajaya*), Mohd Iduan bin Masri (*SK Seri Suria, Kuala Lumpur*) and Nurul Adnin Nadhirah bt Abdul Malik (*SK Datuk Akhir Zaman, Negeri Sembilan*). All of them are primary school teachers.

Part I
Concepts and Contexts

1 Culturally Responsive Science Pedagogy for Social Change and Personal Transformation

Bhaskar Upadhyay

Introduction

History of science mostly shows the value and culture of these disciplines, focusing on objectivity in all aspects of doing and participating in science. So, science contents, ideas, principles, epistemology, and methodologies are all taught as a process of doing and learning science in an objective manner. This push for objectivity has pushed out the value and influence of culture in science and science education. However, many philosophers of science such as Kuhn (1996) have argued convincingly that all scientific disciplines and endeavours are influenced and shaped by the cultures in which the disciplines are practised. Therefore, science is inherently cultural. Since cultures exist in social, historical, and political spheres, it continuously influences how teachers teach science, how and what students want to learn, and how science gets practised and utilized in people's lives. Despite such close connections between science and culture or society, science education and science teaching and learning, in particular, fail to draw from the sociocultural experiences of the community and students to make science learning more fun, relevant, and meaningful. However, there has been much-needed research in science education that has shown the critical value of embedding social and cultural experiences of students in science teaching and learning (e.g. Aikenhead & Jegede, 1999; Albrecht & Upadhyay, 2018; Barton & Upadhyay, 2010; Fortney et al., 2019; Gay, 2010; Ladson-Billings, 1995a; Upadhyay, 2006, 2007; Upadhyay et al., 2020, 2021). In this chapter, the author briefly summarizes and synthesizes the theories and research in culturally relevant/responsive science pedagogy (CRSP) and presents new direction science education needs to take to improve science teacher education programmes, science teaching, and science learning.

Culturally Relevant Pedagogy

The theory of culturally relevant pedagogy (CRP) draws its arguments for culture playing a significant role in teaching and student learning from critical theories, liberation pedagogy, multicultural education, and critical pedagogy. Ladson-Billings (2009) argues that a successful teacher of African American students needs to make oneself familiar with the students' culture and community to make learning

DOI: 10.4324/9781003168706-2

meaningful. She argued that the culture of African American students could not be excluded and considered deficient for learning. From her ethnographic study of successful teachers of African American students, she defined CRP (or culturally relevant teaching) as a "theoretical model that not only addresses student achievement but also helps students to accept and affirm their cultural identity while developing critical perspectives that challenge inequities that schools (and many other institutions) perpetuate" (1995a, p. 469). Therefore, CRP is a way to "empower students intellectually, socially, emotionally, and politically by using cultural references to impart knowledge, skills, and attitudes" (1994, p. 18). CRP clearly asks teachers to take steps to critically examine the "nature of the student-teacher relationship, the curriculum, schooling, and society" as well as "problematize teaching" (Ladson-Billings, 1995a, p. 469) that does not recognize culture as a valuable asset to learning. In the context of teaching students from non-dominant or marginalized (see Upadhyay et al., 2020) groups, the idea of CRP provided both the explanation why these students were failing in schools and the reasons why connections between contents and student culture have positive outcomes for students irrespective of the nature of contents taught. Ladson-Billings proposed three key tenets of CRP that provide the backbone to why CRP supports teaching for social justice, social change, and personal transformation. Students' academic success is central to teaching, followed by teachers' cultural competency with student and community culture. Sociopolitical consciousness is the final tenet of CRP that pushes teachers to continuously make learning that promotes and builds social and political consciousness by leveraging the contents students learn in a class. A key aspect of CRP is that it greatly emphasizes the value in teachers' developing both sociocultural consciousness and a holistic view of caring (Ladson-Billings, 1995b; Morrison et al., 2008).

Culturally Responsive Pedagogy

The culturally responsive pedagogy idea was put forth by Gay (2010) as "using the cultural knowledge, prior experiences, frames of reference, and performance styles of ethnically diverse students to make learning encounters more relevant to and effective for them" (p. 31). Culturally responsive pedagogy or teaching is focused on valuing broad values, beliefs, and knowledge that are based on the racial, cultural, linguistic, and ethnic diversity of students. It has six key components to make teaching culturally connected to students. First, it asserts that teachers need to validate students' cultural heritages to "build bridges of meaningfulness between home and school experiences and lived sociocultural realities" (Gay, 2010, p. 31). Second, it focuses on gaining knowledge of students' culture and the whole child (Gay, 2010) rather than the material contents of a topic – thus supporting students to maintain their cultural identities. Third, it focuses on teaching as a multidimensional practice that "encompasses curriculum content, learning context, classroom climate, student-teacher relationships, instructional techniques, classroom management, and performance assessments" (Gay, 2010, p. 33). Fourth, culturally responsive pedagogy gives students self-determination ability and empowerment. Self-determination and empowerment also help raise higher academic and social expectations in students from underrepresented groups. Fifth, it helps disrupt and push back hegemonic teacher-centred education practices and build more socially and politically conscious classroom activities and structures. Teaching is about educating students to combat racism, sexism, and other forms of oppressive structures. Sixth, it values teaching that makes classrooms and learning more democratic and challenges the idea of universal truths and knowledge permanency. Culturally responsive pedagogy is emancipatory because it "lifts the veil of presumed absolute authority from conceptions of scholarly truth typically taught in schools" (Gay, 2010, p. 38). Thus, culturally responsive pedagogy focuses on what actions teachers can take to make learning more relevant to who the students are and their communities.

In examining Ladson-Billings and Gay's theories, the author finds that both the theories advocate that teachers need to ensure students learn academic concepts and skills at very high standards, build their cultural competency to support students' cultural identities, teach critical reflection and critical analysis skills, and empower students to critique discourses of power and privilege. However, recently CRP has been further expanded and modified to capture the need for cultural sustainability (Paris, 2012; Paris & Alim, 2014) and linguistic revitalization and Indigenous sensitivities (McCarty & Lee, 2014) as core outcomes of teaching and learning.

In brief, culturally sustaining/revitalizing pedagogy or teaching (culturally revitalizing pedagogy) focuses on the value of Indigenous culture, language, and history while teaching Indigenous students in any school but specifically in Indigenous schools. McCarty and Lee define culturally revitalizing pedagogy as "culturally sustaining/revitalizing pedagogy (CSRP) as an approach designed to address the sociohistorical and contemporary contexts of Native American schooling" (McCarty & Lee, 2014, p. 103). They propose three important principles of CSRP if teachers are to productively engage Indigenous students in learning. The Indigenous communities have the freedom to how they want to educate their children. CSRP challenges, disrupts, and reimagines the power imbalance that exists between the dominant group and Indigenous communities. Many indigenous communities have suffered from colonization of their communities. Thus, CSRP intends to change the discriminatory practices inherited from the colonization of their communities in the school curriculum, policies, teaching, and learning. Therefore, countries that have a multitude of Indigenous groups and tribes would highly benefit from CSRP to make teaching and learning more culturally respectful and sustainable.

On the other hand, Paris (2012) and Paris and Alim (2014) proposed that in the dynamic nature of culture and cultural practices (Paris, 2012; Upadhyay et al., 2017), teachers need to think about cultural relevancy in a dynamic manner too. If the culture of a person or a community is an evolving process, then a culturally sustaining idea is more appropriate and more contemporaneous than the idea of cultural relevancy. Therefore, culturally sustaining pedagogy allows teachers and learners to more easily find a balanced space where teaching and learning values continuously and imbeds in everyday classroom practices of critical discourses, questioning, exploratory activities, and assessments. Paris and Alim explain culturally sustaining pedagogy or teaching (CSP)

> seeks to perpetuate and foster linguistic, literate, and cultural pluralism . . . [and] focus[es] on the plural and evolving nature of youth identity and cultural practices and a commitment to embracing youth culture's counter hegemonic potential while maintaining a clear-eyed critique of the ways in which youth culture can also reproduce systemic inequalities.
>
> (Paris & Alim, 2014, p. 85)

A key difference between CSP and culturally relevant/sustaining pedagogy is that CSP accepts that both the historical culture and the new culture that community builds have to be valued and examined in all educational settings. Furthermore, teachers have to play an intentional role in sustaining both the historical culture and the new culture that a student or a community brings.

Examining the key theories and ideas on the value of cultural relevancy in teaching and learning, the author finds that science education needs to unquestionably accept the value of culture in science teaching and learning. The aforementioned theories on cultural relevancy also support the reasons for the poor performance of underrepresented communities in science and the failures of science teacher education programmes, science teachers, policymakers, assessment writers, and curriculum developers to make science a culturally inviting and supporting discipline. In the following pages, the author briefly presents the status of cultural relevancy in science education research, followed by the need for culturally responsive science pedagogy (CRSP).

Status of Cultural Relevancy, Social Change, and Personal Transformation in Science Education Research

Science education research has been regularly exploring cultural relevancy in teaching and learning science for almost three decades in different geographical, economic, racial, gender, and Indigenous contexts. However, most of the work in cultural relevancy has taken place in urban and Indigenous contexts. The major focus of these studies is about how to make science teaching and learning "just about good teaching" (Ladson-Billings, 1995b) that achieves academic success and empowers students to make personal meaning out of science contents in their own lived experiences. The author briefly presents how culturally responsive teaching and learning have addressed the issues of equity, social justice, social change, and personal transformation and empowerment in science education.

Equity and Social Justice

The author believes equity and social justice are cornerstones of anything concerning science education whose major goal is to support and encourage students and teachers from underrepresented groups to succeed in all aspects of science. Equity and social justice challenges in science education are immense across a multitude of groups and communities. One of the ways scholars have attempted to address equity and social justice in science education is through making science teaching and learning culturally relevant to students and their communities. Studies have found that equity and social justice in science education tend to support girls and students from underrepresented groups to both excel in content learning and seek to take actions for social justice (e.g. Barton & Upadhyay, 2010; Upadhyay, 2010). Furthermore, teachers who frame their teaching with equity and social justice as the central agenda for teaching science link issues of equity and social justice that exist in the community with science content. This kind of community connection makes the culture of the students and the communities they represent as valuable assets for engagement. Additionally, the sociocultural nature of equity and social justice issues also more directly encourages students to engage in discourses of community health issues, environment, food security, food deserts, transportation, and racial and linguistic inequities. Therefore, the usefulness of cultural relevance in science is not only about content mastery but also about issues of equity and social justice in the lives of students and teachers.

Social Change and Personal Transformation

Social change and personal transformation are directly linked to engaging and supporting students to notice sociopolitical consciousness and liberation in science content learning and activities (Upadhyay et al., 2020, p. 21). Some scholars of science education (Barton, 1998; Barton et al., 2011; Upadhyay & Albrecht, 2011; Upadhyay et al., 2021) have argued that science for social change and personal transformation is about democratic science that draws from the cultural experiences of students from a multicultural background. Banks's (2004) model of multicultural education also puts sociopolitical and civic actions as the highest dimension of multicultural education. Science education research shows that multiculturalism and pluralistic views in science teaching and learning tends to support students' critical reflection skills and participation in local issues (Atwater, 2010; Jennings & Eichinger, 1999), such as cleaning urban river (Boullion & Gomez, 2001), a voice against racial naming of a disease such as sickle cell disease as a Tharu disease (Upadhyay et al., 2021), soil pollution near a poor community (Morales-Doyle, 2017), and critically examining environmental degradation from the coal power plant near a community (Birmingham et al., 2017), to name a few.

The focus on sociocultural relevancy to improve science teaching in the case of the teachers and various science outcomes for students has produced positive and encouraging results (Aikenhead & Jegede, 1999; Brand et al., 2006; Brown et al., 2018; Chinn, 2002; Dawson, 2014; Mutegi, 2011; Ramos de Robles & Gallard, 2018; Rosebery & Hudicourt-Barnes, 2006; Suriel & Atwater, 2012; Upadhyay & Gifford, 2011; Upadhyay et al., 2017; Upadhyay et al., 2021).

Similarly, research also indicates that greater time spent on critically examining science content with students' lived cultural and social experiences builds a sense of empowerment for personal transformation. Further, when lived experiences are leveraged as valuable funds of knowledge (Moll et al., 1992) to teach science, students have a heightened sense of connection to science and tend to reflect more critically on issues of social change and civic engagement (Barton & Tan, 2009; Upadhyay, 2006; Upadhyay et al., 2017). All of this research indicates to me the need for more focused thinking around what CRSP should look like and why culturally responsive pedagogy would be uniquely different from science. In the following pages, the author proposes the idea of CRSP and its implications to science teaching and learning. However, before moving into sharing aspects of CRSP, the author takes a moment to contextualize the idea of culture in the context of this chapter.

Culture

At the outset, many people think about culture as religion, faith, or something that a community regularly does and on special occasions that is a part of who they are as a group. However, this notion of culture leaves many subtle components embedded in everyday activities that give uniqueness to what and how people do, say, act, and build as a cohesive group. Therefore, culture has varied characteristics and outlooks based on a person's culture and understanding of another culture. For example, a sundried cow dung patty (called *guintha* in Nepali) to a Westerner may look like unhealthy and dirty work, but for a Nepali, it is a source of energy for cooking or keeping warm during cold months. Thus, there is a distinct cultural miss about what an object means to a Westerner versus a Nepali. When culture is described, it tends to represent one of four things: *values, beliefs, norms,* and *expressive symbols* (Peterson, 1979). Another key feature of culture is the *practices* that individuals follow and enact. When these five things are taken together, we find a culture representing many things. Still, every time we think about culture, we observe what people are expressing that makes them distinct from individuals from another culture. However, sometimes culture is less evident to an outside observer because a particular cultural behaviour is subtle. Because culture is an expression, it has multiple markers and meanings depending on the context and local communities. Therefore, I broadly draw from two specific definitions of culture – one is more anthropological, and the second is more critical. According to anthropologist Geertz (1993), culture is "an historically transmitted pattern of meaning embodied in symbols, a system of inherited conceptions expressed in symbolic forms by means of what [people] communicate, perpetuate, and develop their knowledge about and attitudes toward life" (p. 89). On the other hand, critical education theorist Giroux (2001) states:

> Culture is constituted by the relations between different classes and groups bounded by structural forces and material conditions and informed by a

range of experiences mediated, in part, by the power exercised by a dominant society. . . . Culture is constituted as a dialectical instance of power and conflict, rooted in the struggle over both material conditions and the form and content of practical activity.

(p. 163)

The readers and science educators need to understand that culture is a social construct; therefore, culture is dynamic and changes with time and with encounters among people from different cultures. Hence, the author asks science educators to be very cautious when considering culture in their pedagogy and curriculum because each student brings a slightly different culture with them. Stereotypes are built on the notion of culture being static and reducing them to a singular behaviour or characteristic of a particular group. Stereotypes create a discriminatory and demoralizing environment in science teaching and learning spaces. Thus, the author implores science teachers to be more conscious but invites students' culture into science activities and discourses.

Culturally Responsive Science Pedagogy

The author specifically draws from the theories of culturally relevant, responsive, culturally sustaining, and revitalizing pedagogies to form a basis for CRSP. The author then augments CRSP by drawing from critical race theory, critical pedagogy, critical reflection, multicultural education, and democratic practices and values to support further why a framework for CRSP would enhance a more equitable and social justice-oriented science pedagogy. The author proposes CRSP as a framework for teaching science to students from minority groups. In the proposed framework, CRSP is not a method of teaching science but a basis for all science engagements based on the local needs and culture of the students. Next, the author describes critical features of CRSP and the value this brings to science teaching and learning for marginalized students and groups.

Science and Culture in One Place for Success

A goal of CRSP is to bring science and the local cultural assets into everyday teaching for school success. The author specifically focuses on the local because there are many cultural variations within the political boundaries. Cultural variations in some cases overlap with the dominant culture, but in other cases, non-dominant cultures could be peripherally accommodated within the dominant culture. Furthermore, in some instances, non-dominant cultures could be completely ignored and rejected by the dominant culture. Multicultural perspectives (Banks, 2004) include various cultures in teaching and provide equitable opportunities to learn. Furthermore, marginalizing non-dominant cultures from science curriculum and teaching only promotes cursory mentioning of contributions of other cultures in scientific knowledge creation (e.g. Cobern, 1996) and reduces teacher effectiveness (e.g. Brown & Crippen, 2017; Eisenhart, 2000; Upadhyay et al., 2017). When teachers frame their teaching from the point of view of what students are familiar with, learning improves. Congruence between the culture of science and the culture of students always produces increased academic success (Barton & Yang; 2000; Carlone, 2003; Johnson, 2011; Upadhyay & DeFranco, 2008).

Rejection of the Culture of Objectivity

Another feature of CRSP is the rejection of the notion of the culture of objectivity in science and science practices. Science, specifically Western science, has been promoted, practised, and taught as an objective enterprise (Cobern, 1996; Kuhn, 1996; Stanley & Brickhouse, 2001). Modern Western Science (MWS) knowledge is produced, sustained, and disseminated in a manner that looks objective on the surface but the entire scientific enterprise is a White European and Western cultural product. Additionally, the practices of generating questions, hypothesizing, carrying out of the research, and the reporting of the new or rediscovered knowledge are all premised on the culture of no human bias (objectivity). However, it is known that all knowledge, scientific or otherwise, is human-created. Therefore, science cannot be cultureless; rather it is the essence of human culture of exploration and the desire to understand nature and natural systems.

The philosophy of science rejects the idea of the objective nature of observation (Daston, 2008; Feyerabend, 1969; Hempel, 1952; Ogilvie, 2006). The philosophers argue that all observations are mediated by prior human knowledge, culture, instruments, and interpretations. Additionally, the physiology of the human eyes and the instruments used to observe and collect information (data) highly influence what is seen and not seen. Thus, rejecting objectivity in science teaching and learning is important because this then allows learners to see science knowledge as evolving and changing.

Another reason for not requiring learners to believe in scientific objectivity is that it allows for critical reflection (Freire, 1998) on scientific knowledge and the practices of generating it. Even more important is that this values Indigenous and local knowledge of the flora, fauna, and the environment. For example, crop rotation is practised in many Indigenous communities across the globe based on local environment observations, irrigation systems, and fishing net systems developed in the Tharu community in Nepal (Upadhyay et al., 2021); rice-fish coculture in Southern China (Hu et al., 2016), and so on. These are all local practices developed and refined based on centuries of observations and experimentations by the Indigenous and local people without the criteria of objectivity as laid out by MWS. This knowledge is better than lab-based experiments that students learn in many schools without regard to local cultural practices. Knowledge generated by these communities might fail the test of "objectivity" required by science and be considered less "trustworthy or valid", but in the framework of CRSP, this knowledge is at the same level as any knowledge produced by MWS. Therefore, rejecting objectivity and making science more amicable and welcoming to the local expertise would validate local knowledge that works and adds multiple perspectives and critical thinking skills in students.

Social Action and Social Change

An essential feature of CRSP is social action and the desire for social change. Any science teaching and learning engagement must focus on the premise that the learners will gain skills and knowledge to ask questions and generate problems that critically reflect and examine social and cultural discrimination. Critical theory (Giroux, 1983, 1992), critical pedagogy (Freire, 1998; Kincheloe, 2005), and multicultural education (Banks, 2004) advocate for classroom teaching that is more than just content mastery and content knowledge. Students need to become critical consumers of science knowledge to utilize it in situations outside of the school boundaries and within the school boundaries to question and disrupt discriminatory social norms and practices. Socially, economically, and politically science has been considered a discipline that can uplift marginalized groups (NGSS Lead States, 2013) and bring needed changes in communities for a better life (Upadhyay et al., 2020). Yet, in many science classrooms, teachers fail to engage students in examining social issues for action. If students are taught to consider science as facts with very little utility in understanding community issues, problem-posing (Freire, 1998), and challenging dominant narratives of discrimination, then inequities and injustices continue to marginalize people who need the most relief from them. For example, knowledge of Newton's laws of motion could be linked to how and for whose benefit public transportation systems are built (Upadhyay, 2017), how sterilized hybrid seeds displace local seeds and create food insecurity, how the culture of using honey to preserve perishable foods such as berries and fruits supports women's productivity, and so forth. These and many other examples illustrate that science for social change and social actions truly help transform how science should be taught and learned. When thinking about social change and actions in a science class, teachers can easily leverage local culture, practices, and politics. Therefore, CRSP is more than science content and science laboratory experiment skills. CRSP is all about changing the narrative of science from passing the tests to making it actionable for social change through action.

Disruption of Dominant Power

The final feature of CRSP is pushing back and questioning the dominant narrative of science and science culture (Lewis & Aikenhead, 2000). Critical theorists and scholars of education have argued that education at the content level as well as at the institution level needs to focus on how the dominant cultural narratives and practices have created an inequitable and discriminatory environment for learning (Freire,1998; Kincheloe, 2005; Gay, 2010; Giroux, 1993; Ladson-Billings,1995a; McLaren, 2003). The impact of such an environment feels acute when students from non-dominant cultural groups are in science class. Science as a subculture already makes participation in it challenging for students from marginalized and Indigenous communities. For example, the language of science is very different and most of the time alienating even to the dominant group so it is even more foreign to students from marginalized groups. This creates an imbalance in who gets to participate in science and learn to benefit from it. Suppose the goal of science is to empower people on the margins of educational participation; then it would have to equip students to question the subculture of science and the culture of the dominant group that writes the science curriculum and prepares science teachers.

Indigenous science education and other Indigenous scholars have always questioned the culture of science and the culture of the dominant group that establishes the rules of participation in science (Bang & Marin, 2015; Cajete, 1994; Macfarlane et al., 2019; Smith, 1999; Upadhyay et al., 2021). This indicates that many science teachers educated in the dominant subculture of science are less likely to recognize how students from Indigenous cultures and other marginalized cultures struggle to make sense of the science they are learning. Further, the same teachers have had very few opportunities to participate in teacher education programmes that built their teaching skills to challenge the harmful effects of the established culture of a dominant group. If science education is for equity and social justice, we need teachers who are educated in critical theories and critical pedagogies so they are equipped and skilled to teach science that empowers students to disrupt dominant cultural norms of science, curriculum and practices of success that lack cultural relevancy. Therefore, CRSP demands that science teachers engage students in problematizing the culture of science and promoting the culture of learning from multiple cultural perspectives. Thus, CRSP advocates that there is no one right way to do science, and there is no one dominant culture that can dictate how, why, and who students should learn science.

Conclusion

The framework of CRSP has important pedagogical and learning implications. First, it recognizes that students bring a plethora of knowledge and skills that help them make sense of science learning. Students' knowledge and skills are broad and teachers can leverage this knowledge to enhance learning and generate joyful experiences while doing science. Furthermore, the knowledge and skills students bring are locally generated, this means connecting this knowledge to science makes learning more socioculturally relevant and useful.

Second, CRSP encourages teachers to find ways to infuse relevant subcultures of science with the cultural practices of students and their communities. This allows teachers and students to appreciate both the cultures as assets rather than adversaries. Additionally, in science, old knowledge is continuously being revised and new knowledge is added. A similar kind of practice exists in Indigenous cultures, where new experiences in a new environment create new knowledge and help adjust old knowledge. Since people live in a place and they are interacting with the environment, thus what the ancestors learned might need adjustment now. This is what is happening in many Indigenous communities with climate change. Therefore, CRSP can be an excellent framework to support learning science by infusing local ways of generating knowledge.

Third, CRSP is about equity and social justice. Equity and social justice cannot be achieved if the status quo is not questioned and disrupted. CRSP can provide a framework for science teachers to disrupt the dominant narratives in curriculum, teacher preparation, and assessment so that science works more as an asset to bring social change. Furthermore, teachers can utilize local issues (social, political, environmental, racial, etc.) to frame science activities and skills allowing students to critically reflect and seek potential solutions for better outcomes for all. Social change through social activism can be a part of science projects where students utilize current local events or historical events to pose problems that push against the norms that have harmed people in the marginalized groups.

Above all, CRSP allows teachers and teacher educators to intentionally consider science teaching and learning from a culturally responsive and critically problem-posing perspective. The focus of CRSP is to continuously seek to teach and engage students in science that values their culture, language, artefacts, practices, knowledge, and ways of sharing and valuing knowledge for the community's well-being. CRSP is also about ensuring students academically attain success in science so they can pursue the best education for personal and community good. However, CRSP is not about upholding existing social, historical, and cultural inequities just for academic success. As science educators and scholars of equity and social justice, we cannot place the culture of science above the culture of students' community, but we can always find ways to respectfully infuse them in the culture of science. CRSP does not advocate for the wholesale rejection of learning the culture of science, rather it advocates for mutually beneficial ways for the local culture to coexist in science teaching and learning.

Finally, the author ends this chapter by recognizing the challenges of CSRP for many teachers as they have to be willing and flexible enough to recognize the value of culture in science. Furthermore, if the science teacher education programmes do not provide spaces for CRSP in their courses and practicums, teachers will struggle to teach science meaningfully to students who are culturally and linguistically different from them. Science teachers also have to be intentional in their lesson plans, assessments, and daily instruction. How science teachers make CSRP a part of their teaching framework depends on who their students are and how much the teachers believe science is a cultural enterprise. The author believes science teachers are capable of delivering CRSP to their students.

References

Aikenhead, G. S., & Jegede, O. J. (1999). Cross-cultural science education: A cognitive explanation of a cultural phenomenon. *Journal of Research in Science Teaching, 36*(3), 269–287. https://doi.org/10.1002/(SICI)1098-2736(199903)36:3<269::AID-TEA3>3.0.CO;2-T

Albrecht, N., & Upadhyay, B. (2018). What do refugee mothers want in U.S. school science: Learning from the perceptions of science of three Somali mothers. *The Urban Review, 50,* 604–629. https://doi.org/10.1007/s11256-018-0458-9

Atwater, M. M. (2010). Multicultural Science Education and Curriculum Materials. *Science Activities, 47*(4), 103–108. https://doi.org/10.1080/00368121003631652

Bang, M., & Marin, A. (2015). Nature-culture constructs in science learning: Human/non-human agency and intentionality. *Journal for Research in Science Teaching, 52*(4), 530–544.

Banks, J. (2004). Multicultural education: Characteristics and goals. In J. A. Banks & C. A. McGee Banks (Eds.), *Multicultural education: Issues and perspectives* (4th ed., pp. 3–30). John Wiley & Sons.

Barton, A. C. (1998). Margin and center: Intersections of homeless children, science education, and a pedagogy of liberation. *Theory Into Practice, 37*(4), 296–305. https://doi.org/10.1080/00405849809543819

Barton, A. C., Basu, J., Johnson, V., & Tan, E. (2011). Introduction. In J. Basu, A. Calabrese Barton, & E. Tan (Eds.), *Democratic science teaching: Building the expertise to empower low-income minority youth science* (pp. 1–20). Springer.

Barton, A. C., & Tan, E. (2009). Funds of knowledge and discourses and hybrid space. *Journal of Research in Science Teaching, 46*(1), 50–73.

Barton, A. C., & Upadhyay, B. (2010). Teaching and learning science for social justice: Introduction to the special issue. *Equity and Excellence in Education, 43,* 1–5.

Barton, A. C., & Yang, K. (2000). The culture of power and science education: Learning from Miguel. *Journal of Research in Science Teaching, 37*(8), 871–889.

Birmingham, D., Calabrese Barton, A., McDaniel, A., Jones, J., Turner, C., & Rogers, A. (2017). "But the science we do here matters": Youth-authored cases of consequential learning. *Science Education, 101*(5), 818–844. https://doi.org/10.1002/sce.21293

Boullion, L., & Gomez, L. (2001). Connecting school and community partnerships as contextual scaffolds. *Journal of Research in Science Teaching, 38*(8), 899–917.

Brand, B., Glasson, G., & Green, A. (2006). Sociocultural factors influencing students' learning in science and mathematics: An analysis of the perspectives of African American students. *School Science and Mathematics, 106*(5), 228–236. https://doi.org/10.1111/j.1949-8594.2006.tb18081.x

Brown, B. A., Boda, P., Lemmi, C., & Monroe, X. (2018). Moving culturally relevant pedagogy from theory to practice: Exploring teachers' application of culturally relevant education in science and mathematics. *Urban Education, 54*(6), 775–803. https://doi.org/10.1177/0042085918794802

Brown, J. C., & Crippen, K. J. (2017). The knowledge and practices of high school science teachers in pursuit of cultural responsiveness. *Science Education, 101*(1), 99–133. https://doi.org/10.1002/sce.21250

Cajete, G. (1994). *Look to the mountain: An ecology of indigenous education.* Kivaki Press.

Carlone, H. B. (2003). Innovative science within and against a culture of "achievement". *Science Education, 87*(3), 307–328.

Chinn, P. W. U. (2002). Asian and Pacific Islander women scientists and engineers: A narrative explanation of model minority, gender, and racial stereotypes. *Journal of Research in Science Teaching, 39*(4), 302–323. https://doi.org/10.1002/tea.10026

Cobern, W. W. (1996). Worldview theory and conceptual change in science education. *Science Education, 80*(5), 579–610.

Daston, Lorraine. (2008). On Scientific Observation. *Isis, 99,* 97–110.

Dawson, E. (2014). "Not designed for us": How science museums and science centers socially exclude low income, minority ethnic groups. *Science Education, 98*(6), 981–1008. https://doi.org/10.1002/sce.21133

Eisenhart, M. (2000). Boundaries and selves in the making of "science". *Research in Science Education, 30*(1), 43–55.

Feyerabend, P. K. (1969). *Realism, rationalism, and scientific Method* (Philosophical Papers I). Cambridge University Press.

Fortney, B. Morrison, D., Rodriguez, A., & Upadhyay, B. (2019). Rethinking equity in science teacher education: Introduction to special issue. *Cultural Studies of Science Education, 14,* 259–263.

Freire, Paulo. (1998). *Pedagogy of freedom: Ethics, democracy and civic courage.* Rowman and Littlefield.

Gay, G. (2010). *Culturally responsive teaching: Theory, research, and practice.* Teachers College Press.

Geertz, C. (1993). *The interpretation of cultures: Selected essays.* Fontana Press.

Giroux, H. (1983). Theories of reproduction and resistance in the new sociology of education: A critical analysis. *Harvard Educational Review, 53*(3), 257–293.

Giroux, H. (1992). *Border crossings: Cultural workers and the politics of education.* Routledge.

Giroux, H. (1993). Reclaiming the social: Pedagogy, resistance, and politics in celluloid culture. In J. Collins, H. Radner, & A. Preacher Collins (Eds.), *Film theory goes to the movies.* Routledge.

Giroux, H. (2001). *Theory and resistance in education: A pedagogy for the opposition* (2nd ed.). Praeger.

Hempel, C. G. (1952). Fundamentals of concept formation in empirical science. In O. Neurath, R. Carnap, & C. Morris (Eds.), *Foundations of the unity of science* (Vol. 2). University of Chicago Press.

Hu, L. Zhang, J., Ren, W., Guo, L., Cheng, Y., Li, J., Li, K., Zhu, Z., Zhang, J., Luo, S., Cheng, L., Tang, J., & Chen, X. (2016). Can the co-cultivation of rice and fish help sustain rice production? *Scientific Report, 6*(1), 28728. https://doi.org/10.1038/srep28728

Jennings, T. E., & Eichinger, J. (1999). Science education and human rights: Explorations into critical social consciousness and postmodern science instruction. *International Journal of Educational Reform, 8*(1), 37–44. https://doi.org/10.1177/105678799900800105

Johnson, C. C. (2011), The road to culturally relevant science: Exploring how teachers navigate change in pedagogy. *Journal of Research in Science Teaching, 48,* 170–198. https://doi.org/10.1002/tea.20405

Kincheloe, J. L. (2005). *Critical pedagogy primer.* Peter Lang Publishing.

Kuhn, T. S. (1996). *The structure of scientific revolutions* (3rd ed.). University of Chicago Press.

Ladson-Billings, G. (1994). *The dreamkeepers: Successful teaching for African-American students.* Jossey-Bass.

Ladson-Billings, G. (1995a). Toward a theory of culturally relevant pedagogy. *American Educational Research Journal, 32*(3), 465–491.

Ladson-Billings, G. (1995b). But that's just good teaching! The case for culturally relevant pedagogy. *Theory Into Practice, 34*(3), 159–165.

Ladson-Billings, G. (2009). *The dreamkeepers: Successful teachers of African American children.* John Wiley & Sons.

Lewis, B. F., & Aikenhead, G. L. (2000). Introduction: Shifting perspectives from universalism to cross culturalism. *Science Education, 84*, 3–6.

Macfarlane, A., Manning, R., Ataria, J., Macfarlane, S., Derby, M., & Clarke, T. H. (2019). Wetekia kia rere: The potential for place-conscious education approaches to reassure the indigenization of science education in New Zealand settings. *Cultural Studies of Science Education, 14*, 449–464. https://doi.org/10.1007/s11422-019-09923-0

McCarty, T., & Lee, T. (2014). Critical culturally sustaining/revitalizing pedagogy and Indigenous education sovereignty. *Harvard Educational Review, 84*(1), 101–124.

McLaren, P. (2003). *Life in schools: An introduction to critical pedagogy in the foundation of education*. Allyn & Bacon.

Moll, L. C., Amanti, C., Neff, D., & Gonzalez, N. (1992). Funds of knowledge for teaching: Using a qualitative approach to connect homes and classrooms. *Theory Into Practice, 31*(2), 132–141.

Morales-Doyle, D. (2017). Justice-centered science pedagogy: A catalyst for academic achievement and social transformation. *Science Education, 101*(6), 1034–1060. https://doi.org/10.1002/sce.21305

Morrison, K. A., Robbins, H. H., & Rose, D. G. (2008). Operationalizing culturally relevant pedagogy: A synthesis of classroom-based research. *Equity & Excellence in Education, 41*(4), 433–452.

Mutegi, J. W. (2011). The inadequacies of "science for all" and the necessity and nature of a socially transformative curriculum approach for African American science education. *Journal of Research in Science Teaching, 48*(3), 301–316. https://doi.org/10.1002/tea.20410

NGSS Lead States (2013). *Next generation science standards: For states, by states*. The National Academies Press.

Ogilvie, B. W. (2006). *The science of describing: Natural history in renaissance Europe*. University of Chicago Press.

Paris, D. (2012). Culturally sustaining pedagogy: A needed change in stance, terminology, and practice. *Educational Researcher, 41*(3), 93–97. https://doi.org/10.3102/0013189X12441244

Paris, D., & Alim, H. S. (2014). What are we seeking to sustain through culturally sustaining pedagogy? A loving critique forward. *Harvard Educational Review, 84*(1), 85–100.

Peterson, R. A. (1979). Revitalizing the culture concept. *Annual Review of Sociology, 5*, 137–166.

Ramos de Robles, S. L., & Gallard, A. (2018). Sociocultural perspectives: An opportunity to understand science education in a different dimension. In H. Arslan (Ed.), *An introduction to education* (pp. 129–141). Cambridge Scholars Publishing.

Rosebery, A., & Hudicourt-Barnes, J. (2006). Using diversity as a strength in the science classroom: The benefits of science talk. In R. Douglas (Ed.), *Linking science & literacy in the K-8 classroom* (pp. 305–320). NSTA Press.

Smith, L. T. (1999). *Decolonizing methodologies: Research and indigenous peoples*. University of Oxford Press.

Stanley, W. B., & Brickhouse, N. W. (2001). Teaching sciences: The multicultural question revisited. *Science Education, 85*, 35–49.

Suriel, R. L., & Atwater, M. M. (2012). From the contribution to the action approach: White teachers' experiences influencing the development of multicultural science curricula. *Journal of Research in Science Teaching, 49*, 1271–1295. https://doi. org/10.1002/tea.21057

Upadhyay, B. (2006). Using students' lived experiences in an urban science classroom: An elementary school teacher's thinking. *Science Education, 90*, 94–110.

Upadhyay, B. (2007). Developing identity in a new space: Challenges to immigrant mothers in science settings. *Cultural Studies of Science Education, 2*, 500–505.

Upadhyay, B. (2010). Elementary school science teachers' perceptions of social justice: A study of two elementary teachers. *Equity and Excellence in Education, 43*, 56–71.

Upadhyay, B. (November, 2017). *Understanding race and racism in Science through the physics of motion.* Paper presented at the American Anthropological Association Conference, Washington, DC.

Upadhyay, B., & Albrecht, N. (2011). Deliberative democracy in an urban elementary science classroom. In S. Basu, A. Calabrese Barton, & E. Tan (Eds.), *Building the expertise to empower low-income minority youth in science* (pp. 75–83). Sense Publisher.

Upadhyay, B., Atwood, E., & Tharu, B. (2020). Actions for sociopolitical consciousness in a high school science class: A case study of ninth grade class with predominantly indigenous students. *Journal of Research in Science Teaching, 57*(7), 1119–1147. https://doi.org/10.1002/tea.21626

Upadhyay, B., Atwood, E., & Tharu, B. (2021). Antiracist pedagogy in a high school science class: A case of high school science teacher in an Indigenous high school. *Journal of Science Teacher Education, 32*, 518–536. https://doi.org/10.1080/10 46560X.2020.1869886

Upadhyay, B., Coffino, K., Alberts, J., & Rummel, A. (2021). Critical consciousness and empowerment among sixth-grade students in a STEAM classroom: STEAM education for liberation and social transformation. *Asia-Pacific Science Education, 7*(1), 64–95. https://doi.org/10.1163/23641177-bja10020

Upadhyay, B., & DeFranco, C. (2008). Elementary students' retention of environmental science knowledge: Connected science versus direct instruction. *The Journal of Elementary Science Education, 20*, 23–37.

Upadhyay, B., & Gifford, A. (2011). Changing lives: Coteaching immigrant students in a middle school science classroom. In C. Murphy & K. Scantlebury (Eds.), *Coteaching in international contexts: Research and practice* (pp. 267–283). Springer.

Upadhyay, B., Maruyama, G., & Albrecht, N. (2017). Taking an active stance: How urban elementary students connect sociocultural experiences in learning science. *International Journal of Science Education, 39*, 2528–2547.

2 Context Setting

Malaysia, Indonesia and Japan

Nurfaradilla Mohamad Nasri

Introduction

This chapter provides a general overview of the education system practised in each country in terms of its provisions, regulations and levels of organisations. It begins with a brief description of the historical development of each country's education system. The unique demographic scenarios of each country are highlighted throughout the brief description. The languages of instruction for each country are briefly discussed to further analyse the comparison between the three countries. Specifically, this chapter describes science education and the curriculum implemented in each country to better identify their similarities and differences. As the world evolves rapidly, a country's education system progresses concurrently to provide top education to its people.

Malaysia

The Impact of Malaysia's Social and Political Landscape on Its Educational Development

Malaysia is a multi-ethnic country comprising three main ethnic groups: the Malays, who account for half of the Malaysian population (50.4%), followed by the Chinese (23.7%) and Indians (7.1%), while around 10.6% of the population comprises indigenous groups and 8.2% are non-citizens (Department of Statistic Malaysia, 2020). Based on the diversity of the existing Malaysian ethnic groups, it is reasonable to suggest that the Malaysian education system operates within a significantly complex cultural context (Ariffin et al., 2011) and it is crucial that the system is designed to be culturally sensitive and accommodative of the multicultural context.

Traditionally, the Malaysian education system began as a Malay-Islamic-based education before it evolved into a culturally diverse education system during the years of British colonisation. This period of time witnessed massive migration of Chinese and Indian immigrants to the peninsula in order to meet the labour demands in the tin mining and agricultural sectors (Verma, 2002). In comparison to other colonial powers that colonised Malaysia, the British initially established

DOI: 10.4324/9781003168706-3

and introduced their own schooling system to provide a minimum level of knowledge and skill acquisition to the people. On that note, Saad (1980) reported that there were three primary goals of education during the British colonial period, namely (1) Socialising children with foreign values, attitudes, norms and cultures and not the native values, attitudes, norms and cultures; (2) appointment of native community leaders to meet their specific goals in gaining support and trust from the local people; and (3) producing the lowest-level workforce to be engaged directly or indirectly in the agriculture sector. Therefore, it is clear from these three goals that the purpose of introducing the so-called systematic education system in Malaya by the British was primarily for the benefit of the British people themselves economically and politically.

During the British administration, there were four types of schooling systems, namely (a) Malay Vernacular School, (b) Chinese Vernacular School, (c) Tamil Vernacular School and (d) English Schools. Despite gaining independence from the British for more than five decades, the current Malaysian education system still carries significant influence or values from the previous British education system.

The country's compulsory school system is divided into two major parts, namely primary education (*sekolah rendah*) for six years beginning at the age of seven years old and ends with a national assessment examination in year six. Secondary education comprises three years of lower secondary education with a national examination (known as *Pentaksiran Tingkatan 3*- PT3) at the end of the third year, while the upper secondary education spans over two years which ends with a national exam (known as *Sijil Pelajaran Malaysia* – SPM) that is equivalent to the ordinary level (O-level) in the English system. It is important to note that the schooling system at this level of education varies and is racially driven based on the national and vernacular schools. Although heated arguments and debates continue to take place around this issue, the success of this system in providing basic education to students regardless of racial backgrounds or identities should be greatly highlighted. Next, upon completion of the compulsory education, several optional pathways exist for entrance into tertiary education, ranging from a minimum of 1.5 years of pre-university education/matriculation to a higher school certificate (known as *Sijil Tinggi Persekolahan Malaysia* – STPM) which is comparable to the advanced level (A-level) in the British education system.

In general, the primary purpose of education in Malaysia is to produce a holistic individual who can contribute to national prosperity and unity. This aspiration is clearly expressed in the statement of the National Education Philosophy (NEP), where the significant emphasis is given to developing the physical, emotional, spiritual and intellectual aspects of an individual. Guided or framed by the NEP, the education system is geared and moulded with Islamic values that greatly emphasise on producing good human beings.

Overview of Malaysia's Education System

Education in Malaysia is among the top priorities of the federal government. The government is committed to providing quality and equitable education to all children in the country regardless of gender, race and social class. The government's serious commitment to education is clearly evident in the country's five-year development plan – the Malaysia Plan, which emphasises the importance of access to quality education, effective training programmes and lifelong learning.

Among the significant changes in the education system in Malaysia are the use of English in the teaching of science and mathematics through the Dual Language Programme (DLP), emphasis on providing quality STEM education to increase student enrolment, attention given to ensuring access to quality and equitable education for all children, emphasis on celebrating inclusive education by encouraging the placement of special needs students in mainstream classes and the transition from an exam-oriented education system to one where the education focuses on providing meaningful as well as deep learning experiences for its students.

Malaysia provides 11 years of free primary and secondary education (grades 1 to 11). Primary schooling (grades 1 to 6) is compulsory for all children. It is divided into two levels: Level 1 (grades 1 to 3) and Level 2 (grades 4 to 6). Upon completion of the lower secondary education (grades 7 to 9), students continue their schooling at the upper secondary level (grades 10 to 11) in the arts, sciences or technical and vocational streams. After completing grade 11, students may opt to enrol in grade 12 (the first of two final years of schooling typically taken by students who intend to enrol in public universities in Malaysia); pre-university foundation programmes (matriculation programmes offered at private universities in Malaysia); or skill, technical and vocational programmes.

Malaysia has long practised an education with a centralised curriculum that focuses on developing holistic and well-balanced individuals. Primary education in Malaysia is aimed at producing a balanced, harmonious and high-morale individual through an integrated development of the child's potential in terms of the intellectual, spiritual emotional as well as physical aspects. The primary education curriculum comprises a wide range of subjects, including languages, arts, sciences, Islamic and moral education, and design and technology. Meanwhile, the secondary education curriculum includes comprehensive education programmes that serve as a bridge to tertiary education. Secondary education is divided into two cycles, namely the lower secondary which comprises Forms I–III and upper secondary which includes Forms IV and V. Being an extension to the primary level, the aim of secondary education is to further develop the child's potential in an integrated, balanced as well as holistic manner. Upon completion of the three-year lower secondary programme, students are offered academic, technical or vocational tracks at the upper secondary level. Covering a period of two years, students in academic and technical tracks sit for the Malaysian Certificate of Education (MCE) Examination, while those in vocational tracks sit for the Malaysian Certificate of Education (Vocational) Examination. Finally, the tertiary education or post-secondary education in Malaysia can be divided into Form VI programmes that prepare students for the Malaysian Higher School Certificate (STPM) Examination and matriculation courses that prepare students for matriculation examinations. Tertiary education providers or higher education institutions aim to produce employable graduates to meet the constant need for human resource capital for the country's economic growth.

Languages of Instruction

Malaysia is a multi-ethnic country with a population of 32.7 million in which three major ethnicities – Malay, Chinese and Indian – make up the majority of the population (Department of Statistics Malaysia, 2020). Consequently, the common language used in national schools is *Bahasa Melayu* or the Malay Language which is also the national language of the country, while vernacular schools use either Mandarin or Tamil. It is important to note that despite the different languages of instruction, both types of schools implement the Malaysian National Curriculum.

Malaysia's Science Curriculum

The science component in the national curriculum was designed to provide students the opportunities to acquire scientific knowledge and skills, develop critical and creative thinking skills and apply their knowledge and skills in everyday life. The curriculum stipulates that learning activities should serve the purpose of stimulating students' critical and creative thinking skills and should not confine students to routine or rote learning practices. At this level, the curriculum is organised into three domains: scientific knowledge, skills, and scientific attitudes and values.

- Scientific Knowledge – This domain encompasses interrelated concepts, facts, rules and principles associated with biological, chemical and physical processes as well as astronomy and technology.
- Skills – Scientific skills are important in scientific investigations such as conducting experiments and carrying out projects.
- Scientific Attitudes and Values – Scientific attitudes and noble values are instilled through experiential learning either spontaneously or through planned activities.

Instructional Science Practices in Malaysia

Formerly, a traditional instructional practice that emphasises teacher-led teaching was fully utilised in the Malaysian school. This teaching approach incorporates repetitive learning styles through one-way didactic communication with great attention given towards preparing students for high-stakes standardised examinations (Tengku Kasim, 2014). The instructional science practices in Malaysia are largely characterised by the adoption of certain teaching and learning techniques such as lecturing, rote learning and spoon-feeding (Tengku Kasim, 2014). According to Tengku Zainal et al. (2009), some Malaysian teachers prefer to continue teaching using the traditional practices as opposed to being creative or inventive in their teaching approach.

Saleh and Aziz (2012) described Malaysia's instructional practice as having a negligible level of interaction as most of the talking and instruction are carried out by the teachers while only a few students would contribute their views. Occasionally, the teachers would conduct demonstrations and laboratory activities to verify the concepts taught in the classroom and provide some exercises which are given at the end of the textbook to familiarise students with examination questions (Saleh & Yakob, 2014; Saleh & Aziz, 2012). This description is in line with Li and Arshad's (2015) findings where the majority of Malaysian science teachers in their study were found to conduct traditional instructional practices by giving clarification theoretically in their lessons.

The heavy reliance of Malaysian teachers on the aforementioned science instructional practices is most likely due to the overt emphasis on exams in the Malaysian education system. Moreover, the emphasis on high-stakes national examination to some extent causes most science teachers to mainly focus on excellent-performing students while ignoring the weaker ones (Abdullah Mubarak & Abdul Razak, 2017).

Saleh and Liew (2018) recognised that although an increasing number of Malaysian teachers are practising student-centred teaching approaches, most of them are still comfortable playing the dominant role in managing their classrooms. Malaysia's performance in Programme for International Student Assessment (PISA) revealed that almost half of Malaysian students failed to achieve the minimum level of scientific literacy and to make it worse, less than 1% managed to achieve the highest level of performance. Among the science teaching strategies proposed by the MOE are (1) inquiry-discovery approach which emphasises learning through exploratory learning experience; (2) constructivism approach which suggests that students learn best by building their own understanding; (3) science, technology and society approach which recommends science learning to be taught through investigation and discussion with the primary aim of solving issues and problems related to science, technology and society; and (4) contextual approach which focuses on ensuring the connectivity between formal science learning to students' daily life activities.

In relation to the core concept of this book – culturally responsive science pedagogy – it is important to highlight that Malaysia is showing significant progress in evolving into a relatively united nation of diverse ethnic groups; hence, the knowledge about the customs and cultures of the ethnic groups in Malaysia is given priority and is introduced to students as one of the elements to be taught across the curriculum. However, the effort to link science teaching with students' background beyond the elements of ethnicity (e.g. socioeconomic, geographical location, religion) is relatively limited. For instance, Alt (2018) reported that very few science teachers are capable of using community resources, the school environment and the economic activities of the local community as science teaching examples in the formal classroom. As a result, many students reported that formal science learning is irrelevant and does not make sense in their life. According to Darling-Hammond et al. (2020), students whose prior knowledge is not used during science learning could not apply what they have learnt in the classroom to solve their daily life problems. Although various efforts have been made by the Ministry of Education (MOE) to encourage teachers to link science teaching with students' daily lives, this aspect is only generalised to represent all Malaysian students and does not refer to the unique sociocultural and geographical elements of the individual schools.

Indonesia

The Impact of Indonesia's Social and Political Landscape on Its Educational Development

Indonesia, or officially known as the Republic of Indonesia, is one of the South-East Asian countries that has a large geographical size spanning over 17,000 islands. Indonesia is home to hundreds of different ethnic groups and cultures. However, more than half of the population can be mainly classified as belonging to two main ethnic groups – Javanese (41% of the total population) and Sundanese (15% of the total population).

Education in Indonesia began in the early kingdom era or also known as the Hindu-Buddhist civilisation period. The education system during this era is known by the term *karsyan* which means a place of the hermitage. However, the emergence of Islamic state in Indonesia led to the acculturation of both the Islamic tradition and Hindu-Buddhist tradition. The *Pondok Pesantren* which is a type of Islamic boarding school was established around this time and its location resembled the location of *karsyan*. The emphasis in *Pondok Pesantren* is on inculcating core values of sincerity, simplicity, individual autonomy, solidarity and self-control, where the aim is to deepen students' knowledge of the Quran, particularly through the study of Arabic, traditions of exegesis, the Sayings of the Prophet (Hadith), law and logic.

During the Dutch colonial era, elementary education was introduced following the Dutch education system which was based on the social status of the colony's population. The best education institutions were only available and reserved for the European population. However, some of these exclusive Dutch-founded schools later opened their doors for native Indonesians and were remodelled as *Sekolah Rakyat* or *Sekolah Dasar*. For native Indonesians who lived in rural areas, the Dutch built the Desa Schools or village school system to help increase literacy among that particular group of the population with lesser funds than the European schools.

During World War II, Indonesia was occupied by the Japanese. As a result, the existing multi-operational Dutch education system was consolidated into one single operation to reflect the Japanese education system. The Japanese occupation led to a newer form of education which was anti-West oriented and thus included the indoctrination of Japanese culture and history. The primary aim of education was to create the Greater East Asia Co-prosperity Sphere of Influence. After the country got independence in 1945, the education system was revolutionised as one that is anti-discriminatory, anti-elitist and anti-capitalist with greater emphasis on religious knowledge.

Overview of Indonesia's Education System

Education at the national level in Indonesia is based on *Pancasila* – the philosophical foundation of the Indonesian state set forth in the 1945 Constitution of the Republic of Indonesia and enacted in June 2003 (Law Number 20). Focusing on producing a holistic individual, the role of education in Indonesia is to establish a civilised-state as well as to ensure the nation's economic growth and development by enhancing its students' intellectual capacity and fostering certain human values such as being faithful and pious to God, possessing a high-moral standard and noble character, being physically and mentally healthy, well-informed, creative thinkers and independent individuals, and being democratically responsible citizens.

In accordance with Government Regulation Number 25 (2000), there are some aspects primarily provisioned by the Ministry of National Education, namely student competency standards, the national curricula and national assessments; textbook; the acquisition of academic degrees; national education budgeting; and instructional hours for both the primary and secondary education levels.

Law Number 23 (2014) classifies governmental relations into three categories: absolute national authorities, concurrent national and regional authorities and presidential authorities. The national authorities manage teacher education and provide teacher professional development programmes through national education standards. In contrast, the regional authorities are responsible for transferring teachers within provinces or districts while managing basic education schools and non-formal schools at the same time.

Indonesia's education system consists of multiple streams involving formal education, non-formal education and informal education. These streams are designed to complement and enrich each other mainly through face-to-face classroom interactions, which may also be supplemented or substituted with distance learning at higher levels of education. Formal education comprises three levels – basic, secondary and higher education – covering several knowledge disciplines – general, vocational, professional, vocational-technical, religious and special education. All streams, levels, and types of education are educational units organised by the national government, local governments, the community or any combination thereof.

Languages of Instruction

Indonesia has a large population that includes many diverse ethnic groups, resulting in the practices of various languages, and among the most prominent languages used are Javanese and Sumatran. Although over 400 languages are spoken in Indonesia, almost 60% of the population is competent in the Indonesian national language known as *Bahasa Indonesia* which was declared as the state language in the 1945 constitution. Being the official national language of Indonesia, *Bahasa Indonesia* is the primary language of instruction in schools. *Bahasa Indonesia* not only serves to promote unity, but it also fulfils four main functions which are cognitive, instrumental, integrative and cultural. English is introduced later in grade 7. However, English is the language of instruction in international schools with their respective international curriculum standards.

Indonesia Science Curriculum

Indonesia introduced and developed a new national curriculum, K13, in 2013. Beginning in July 2013, 6% of schools implemented the new curriculum in grades 1, 4, 7 and 10. The K13 curriculum includes a significant improvement for science instruction (Mukminin et al., 2019). Prior to the senior secondary level, all elementary and junior secondary school students learn basic science. However, at these levels, science is taught as an integrated and thematic subject with no distinct separation between physics, chemistry and biology content. In addition to acquiring science knowledge, K13 emphasises on promoting students to work scientifically and develop scientific attitudes. The new curriculum includes the following scientific behaviours or skill objectives: observing, questioning, exploring, associating and communicating; all these skills are essential for developing twenty-first-century skills. Science instruction is delivered in local and global contexts.

Japan

The Impact of Japan's Social and Political Landscape on Its Educational Development

The sixth century marked the beginning of formal education in Japan which was characterised by the adoption of Chinese culture and integration of both Buddhist and Confucian teachings. Despite having a systematically structured education system, scholar officials at the court were not chosen through an Imperial examination system; instead, it remained hereditary family possessions. The influence of scholar officials ended with the rise of the military class during the Edo period. During this period, the country was largely pacified and thus rather than competing through war to increase their influence, the military class turned to the economic field. Therefore, their warrior-turned-bureaucrat Samurai elites were educated not only in military strategies and the martial arts but were introduced to agriculture techniques and accounting skills.

Following World War II, efforts to eradicate militarist teachings and promote democracy in Japan by the allied occupation government gained pace. Correspondingly, an education reform was initiated based on the American model of education which included lifting the burden of entrance examinations, promoting internationalisation, leveraging educational technologies, diversifying education and supporting lifelong learning.

Overview of Japan's Education System

The Fundamental Law of Education which formed the basis of the post–World War II education in Japan was enacted in 1947 and amended in 2006. This law established the basic principles of Japanese education by mainly focusing on providing students with equal opportunities to receive free and compulsory education for nine years. More importantly, it served as the foundation of all education-related laws in Japan, including the School Education Law and the Social Education Law.

Under Japan's curricular reform, the national curricula or known as the Courses of Study have been constantly revised since their implementation in 1947. Educational reform is essential to responsively adapt to the nation's needs and to keep up with significant societal changes over the years. Most importantly, the education reform in Japan occurs primarily due to the students' changing needs of learning for each age group. The revised Course of Study for Elementary Schools was announced in March 2008 and was fully implemented in April 2011. Some parts of the new curricula for science were implemented partially during the transition period from April 2009 to March 2011. Meanwhile, the revised Course of Study for Lower Secondary Schools was announced in March 2008 and was also fully implemented in April 2012; however, some parts of the new curricula for science were implemented partially during the transition period from April 2009 to March 2012.

For educational administrative purposes, Japan set up the Ministry of Education, Culture, Sports, Science, and Technology (MEXT) to regulate school education, and all educational activities in Japan come under the supervision of MEXT. Moreover, MEXT supervises and subsidises local boards of education in order to allow optimum level of operations.

A point to note is that Japan, like Malaysia and Indonesia, also has both public and private education institutions that exist at all levels of the academic organisation level. However, only 5% of Japan's schools are private, and international schools are only available in large cities. In terms of education funding, the federal government bears most of the expenses for national schools, while municipal and prefectural schools are supported locally, with some assistance from the federal government. In contrast, private schools are self-supporting through tuition, donations and contributions from businesses.

Education in Japan follows a 6-3-3 pattern: six years of primary school, three years of lower secondary school and three years of upper secondary school. Some students attend six years of secondary school which combine lower secondary education with both general and specialised upper secondary education. Introduced into the school system in April 1999, these comprehensive secondary schools are designed to focus on the diverse needs of secondary school students. Of students attending these comprehensive schools, 25.1% are enrolled in private six-year secondary schools.

Japan's compulsory education consists of six years of primary education and three years of lower secondary education, and almost all children ages 6 to 15 are enrolled in school. In upper secondary schools, education can be full-time, part-time or correspondence. Full-time students can complete upper secondary school within three years, while part-time and correspondence students might take a longer period. Although high school education in Japan is not compulsory, approximately 98% of students choose to continue their education by attending high school. Having said that, a great majority of Japanese students either go to university, technical college, trade school or junior college, or find employment after graduating from their high school.

Aiming to promote students' interest in education and ensuring learning experiences that are driven by students' curiosity about the world, there is no official policy on educational streaming, and students are not tracked in public primary and lower secondary schools. From primary to the end of lower secondary school, a compulsory programme of science is taught to all students in mixed ability classes to ensure all Japanese students acquire at least a minimal level of science literacy. However, beginning in grade 7, schools may offer several optional subjects for interested students. On the other hand, at the upper secondary level, schools place students into tracks according to their entrance examination results, offering courses that are geared towards differing abilities and interests that serve to address the learning needs of all students. After going through ten years of common and standardised science curriculum, schools offer several different curriculum options in science in grades 11 and 12. Upper secondary education is divided into two main streams: general secondary education which provides general academic preparation and specialised secondary education which provides vocational as well as other courses that are designed for students who are preparing for a specific career.

Japan is widely known for its typical teaching and school culture, Japanese teachers' perspective or way of thinking in relation to pedagogical concepts and the struggles of teachers to carry out learning study or learning community which is well known as 'lesson study by concentrating on how students learn rather than how teaching and learning should be delivered, teachers in Japan focus on nurturing the thinking ability of students' (Itō & Nakayama, 2014). These cultures have been strongly argued as a stimulus to the quality of students and teachers in Japan and the high performance in PISA. As Japan is riding a new wave of cultural diversity, Japan emphasises the policy of equity and hence efforts are actively being made to ensure equitable education in this country.

Languages of Instruction

According to the 2013 population data, Japan's population stood at approximately 127.3 million. The majority of Japan's inhabitants are Japanese. Recently, foreign national residents (medium term to long term residents and special permanent residents) have gradually increased in number. As Japanese is spoken by the overwhelming majority of Japanese people, Japanese is used as the primary language of instruction in all Japanese schools. However, in some regions where there are Brazilian communities, the education is provided both in Japanese and Portuguese. This diversity in the community offers great potential for offering a culturally responsive education for all students.

Japanese Science Curriculum

Science instruction begins in the grade 3, and science is a required subject throughout compulsory education. The science curriculum consists of three parts: overall objectives for the level (primary, lower secondary or upper secondary), objectives and content for each grade or section, and syllabus design. All schools in Japan are required to address all components of science instruction by formulating an overall plan for science that includes descriptions of the following: objectives and content; qualities, abilities and attitudes to be fostered; learning activities; teaching methodology and teaching framework; and the evaluation of learning.

Two noticeable characteristics of science education in Japan are their comfortable, neat and highly equipped science laboratories. All science laboratories in Japanese schools share similar features where they are equipped with the same apparatus and technology. This is because of the science education promotion law which enforces and regulates the standards of science teaching aids and supports its expense by the national treasury.

Another notable characteristic of Japanese science education is the school science garden at the elementary level. Focusing on delivering science education through experiential learning approaches, it is necessary for each Japanese school to have their own science garden which may assist student's pursuit of learning about living things and their lives, and the earth and universe. Since the Japanese school year starts in early April and ends in late March, they can experience seeding to fruition seamlessly. Science gardens play a distinctive role particularly among urban students who seldom have any interaction with nature.

Conclusion

Critical analysis of the literature revealed similarities and differences between Malaysia, Indonesia and Japan's education systems. These three countries share or operate under a centralised education system which is characterised by an authoritative control of the federal government over the curriculum, teacher training, textbook, assessment as well as educational funding. However, unlike Malaysia, both Indonesia and Japan grant a certain degree of authority to state and district educational bodies. As a country that was once colonised by several world powers, the education system in Malaysia and Indonesia during the early stages was intended to guarantee the occupation of the colonialists and ensure that the indigenous people remained illiterate. Additionally, the large influx of immigrants which were brought in by the colonisers into Malaysia and Indonesia for the purpose of economic growth and development resulted in a society that has unique ethnic diversities, which later became an ongoing challenge for both Malaysia and Indonesia in uniting their communities after independence. In contrast, Japan which initially remained and preserved its homogenous culture faces an emerging phenomenon of an expanding ethnically diverse population. Despite all countries attaining their own national curriculum to ensure all citizens receive the same quality education with common shared values, the language used. which is also the national language of the respective countries, may pose a risk towards marginalising the minority ethnic groups. Therefore, respecting the concept of equitable education, Japan has been actively developing its education system to serve as an effective platform for addressing the issue of diversity by meticulously designing its education approach to be more sensitive in meeting the needs of each student from different cultural backgrounds, which in the context of this book refers to culturally responsive education.

References

Abdullah Mubarak, A., & Abdul Razak, N. (2017). Malaysian students' achievement in TIMSS 2011: Does Science inquiry really matter? *Malaysian Journal of Learning and Instruction (MJLI)*. Special issue on Graduate Students Research on Education, 1–25.

Alt, D. (2018). Science teachers' conceptions of teaching and learning, ICT efficacy, ICT professional development and ICT practices enacted in their classrooms. *Teaching and Teacher Education, 73*, 141–150.

Ariffin, S. R., Daud, F., Ariffin, R., Rashid, N. A., & Badib, A. (2011). Profile of creativity and innovation among higher learning institution students in Malaysia. *World Applied Sciences Journal 15 (Innovation and Pedagogy for Lifelong Learning)*, 36–41.

Darling-Hammond, L., Flook, L., Cook-Harvey, C., Barron, B., & Osher, D. (2020). Implications for educational practice of the science of learning and development. *Applied Developmental Science, 24*(2), 97–140.

Department of Statistic Malaysia. (2020). www.dosm.gov.my/v1/index.php?r= column/cthemeByCat&cat=430&bul_id=Szk0WjBlWHVTV2V1cGxqQ1hyVlpp Zz09&menu_id=L0pheU43NWJwRWVSZklWdzQ4TlhUUT09

Itō, Y., & Nakayama, S. (2014). Education for sustainable development to nurture sensibility and creativity: An interdisciplinary approach based on collaboration between "Kateika" (Japanese Home Economics), Art, and Music Departments in a Japanese Primary School. *International Journal of Development Education and Global Learning, 6*(2), 5–25. https://doi.org/10.18546/IJDEGL.06.2.02

Li, W. S. S., & Arshad, M. Y. (2015). Inquiry practices in Malaysian secondary classroom and model of inquiry teaching based on verbal interaction. *Malaysian Journal of Learning and Instruction, 12*, 151–175.

Mukminin, A., Habibi, A., Prasojo, L. D., Idi, A., & Hamidah, A. (2019). Curriculum reform in Indonesia: Moving from an exclusive to inclusive curriculum. *CEPS Journal, 9*(2), 53–72.

Saad, I. (1980). *Competing identities in a plural society: The case of peninsular Malaysia*. Institute of Southeast Asian Studies.

Saleh, S., & Aziz, A. (2012). Teaching practices among secondary school teachers in Malaysia. *International Proceedings of Economics Development and Research, 47*(14), 63–67.

Saleh, S., & Liew, S. S. (2018). Classroom pedagogy in German and Malaysian secondary school: A comparative study. *Asia Pacific Journal of Educators and Education, 33*, 57–73. https://doi.org/10.21315/apjee2018.33.5

Saleh, S., & Yakob, N. (2014). Teachers' conceptions about physics instruction: A case study in Malaysian Schools. *Australian Journal of Basic and Applied Sciences*, *8*(24), 340–347.

Tengku Kasim, T. S. A. (2014). Teaching paradigms: An analysis of traditional and student-centred approaches. *Journal of Usuluddin*, *40*, 199–218.

Tengku Zainal, T. Z., Mustapha, R., & Habib, A. R. (2009). Pengetahuan pedagogi isi kandungan guru Matematik bagi tajuk pecahan: Kajian kes di sekolah rendah (Content pedagogical knowledge of Maths teachers for the topic of fractions: Case study in primary schools). *Jurnal Pendidikan Malaysia*, *34*(1), 131–153.

Verma, V. (2002). *Malaysia, state and civil society in transition*. Lynne Rienner Publishers Inc.

3 Equitable Science Education through Culturally Responsive Science Pedagogy

Nurfaradilla Mohamad Nasri

Introduction

Throughout history, scientific breakthroughs and revolutions have underpinned substantial economic growth. Science-based human capital is among the key driving factors for innovation and technology. However, science education is facing another stumbling block where students' interest to continue studying in science-related subjects has been observed to decline. Students have been found uninterested in learning science-related subjects due to various reasons. Several approaches have been made to find a solution for this issue. One of the approaches in solving this issue is to ensure that their learning experiences are meaningful. Responding to this complex issue, culturally responsive science pedagogy (CRSP) has been introduced to address educational equity challenges along the lines of urban–rural geographical locations, ethnicity, gender, socio-economic status and other dimensions.

What Is Educational Equity?

Many developing countries reported that they are struggling to provide equitable opportunities for all students to learn science. Their struggles were evidenced by persistently lower levels of students' science proficiency in the international assessments such as Programme for International Student Assessment (PISA) and Trends in Mathematics and Science Studies (TIMSS). Although Malaysia through its strategic plan of 12 years aspires to improve both quality and equity in education, nonetheless, the disparities in the allocation of educational resources and teacher supply between rural and urban schools remain a major problem. This poorly managed systemic problem has led to uneven student educational outcomes, where students from urban schools are more likely to perform better compared to their disadvantaged counterparts. For example, in the 2018 Primary School Achievement examination (UPSR), the gap between urban and rural schools was five percentage points in favour of urban schools, while in Malaysian Certificate of Education examination (SPM), the gap was two percentage points. Yet, this unresolved problem is far more critical than just the location of the school as education

DOI: 10.4324/9781003168706-4

equity prevails in the aspect of gender, where the academic achievement gap between boys and girls in Malaysia is very large with girls leading in most science-related subjects (Goolamally & Ahmad, 2010).

Although there has been a great deal of research carried out to investigate and propose different ways of helping us to understand and promote educational equity, it remains a very much contested educational concept (Hodge, 2020). Furthermore, the elusive concept of equity has led the term of equity and equality to be used interchangeably in the literature, creating some confusion about the concept of equity. Equality focuses on ensuring all students receive the same amounts of support irrespective of individual needs (Barrance & Elwood, 2018). In contrast, equity which implies the "idea of need" purposefully distributes unequal support according to individual needs (Ramírez & Pacheco, 2016). Undoubtedly, all students should have equal access to high-quality education; however, being an educator, we are morally and ethically responsible to ensure that our disadvantaged and underserved students receive more support. On that note, it could be argued that equity which refers to the process of ensuring all students receive resources that they need denotes "more for those in need to help them move up" while equality refers to the equal opportunity to success.

Equity in education can be defined in many different ways. According to Field et al. (2007), equity in education comprises two aspects which are inclusion and fairness. Based on this point of view, they asserted that equity as inclusion means ensuring all students reach at least a basic standard level of knowledge and skills to function effectively as an active, effective and responsible citizen. In contrast, equity as fairness means ensuring that students' socio-economic status, cultural and demographic background as well as their personal characteristics do not hinder them from reaching their full academic potential. Similarly, Howard (2013) pointed that equity in education does not mean giving everything or giving the same things to all students. In the same vein, Hodge (2020) defines equity as an assurance which guarantees that all students have equitable access to resources that they need despite their background.

Perhaps the most thorough and comprehensive definition of equity is given by Nasir et al. (2006, p. 499) who explained that equity is "not about offering or producing sameness," but about ensuring that all students can "live the richest life possible and reach their full academic potential". Based on the literature reviewed, it is clear that in the context of education, equity focuses on reducing the educational gap between the top and bottom by allocating resources according to the students' needs. Informed by researchers' definitions of educational equity, the author opts to define equity in education as an initiative to provide all students with adequate opportunities to learn and expect them to meet the highest academic standards. In the context of science education, we would define educational equity as a systemic initiative in providing high-quality science learning opportunities to all students while expecting them to achieve the highest science academic standards.

Why the Need to Emphasise on Educational Equity in Science Education?

Science education is vital for producing an educated workforce to drive a country's economic growth, and, most importantly, it also is the key to better social outcomes for individuals. However, focusing on delivering high-quality science education through equal access alone is not enough; greater attention should be put to address equity in science education. Equity in science education is important for several reasons:

1   An equitable science education guarantees inclusion and fairness for all students by levelling the playing field which helps the students to develop their scientific knowledge and skills (Ainscow, 2020)
2   Equitable science education not only supports and ensures the success of all students in science learning, but most importantly, it gives underserved students the lift that they need to thrive (Avendano et al., 2019)
3   Equitable science education which provides opportunity to all students to develop their full potential and capacities in science education echoes the human right for quality education applauded in the United Nations Declaration on the Rights of the Child (Barrance & Elwood, 2018)

In conclusion, educational equity thus has a range of benefits especially in science education as it paves the way for academic and social excellence which is crucial to drive nations' development.

Current Status of Educational Equity in Science Education: Malaysian Context

One of the biggest problems facing the Malaysian education system is the significant political interference in determining the national education policies with an eye on re-election. Hence, unsurprisingly, education in Malaysia faces constant educational turbulence and is rife with serious tensions between (a) short-term objectives and long-term aspirations, (b) local identity and global demand, (c) knowledge transmission and knowledge creation, (d) informative-dense curriculum and innovative-responsive curriculum, (e) assessment for selection and assessment for development and so on (Osman & Marimuthu, 2010).

Believing that a student's education success should not be determined by their background, Malaysia is on a serious mission to end educational inequity. Various strategic planning was highlighted in the Malaysia Education Blueprint 2013–2025 (Preschool to Post-Secondary Education), the Malaysia Education Blueprint 2015–2025 (Higher Education), the Government Transformation Programme, the Eleventh Malaysia Plan 2016–2020: Anchoring growth on people, the Economic Transformation Programme, and the Malaysia Education for All (End Decade Review Report 2000–2015), to raise the bar of achieving quality education for ALL.

In terms of access to education, Malaysia can be proud when in 2017, 97.9% of Malaysian children were reported educated at the primary level. Enrolment rates at the lower secondary level (Form 1 to Form 3) had risen from 95.0% in 2016 to 96.6% in 2017. Meanwhile, for the upper secondary level (Form 4 and Form 5), enrolment rates increased by 0.8%, from 85.8% in 2016 to 86.6% in 2017 (Department of Statistics Malaysia, 2018). Although these enrolment rates are slightly lower than that of high-performing education systems like Singapore, the rates are nevertheless significantly higher than most developing countries.

Unfortunately, despite significant improvement in access to education and various deliberate efforts to raise the quality and equity of education in Malaysia, the achievement gap among Malaysian students still prevails. For instance, even though Malaysia can be proud of the increase of 0.6% in students achieving "A" in the 2017 SPM English language subject, the passing rate declined by 0.3% from 79.4% in 2016 to 79.1% in 2017. This situation indicates that there are some flaws in Malaysia's education system.

Furthermore, the performance of Malaysian students in PISA in 2016 indicates that the basic education outcome in Malaysia is below its regional peers. For instance, Malaysia was ranked 22nd and 24th for mathematics and science respectively in a list of 39 participating countries. This alarming result raised questions of what may have gone wrong or what aspects may have been overlooked.

Many researchers have proposed a great number of systemic and societal factors that contribute to educational inequality particularly in science education. However, a close examination of these factors revealed that researchers' personal perception of one's educational success remarkably affects their ideas about the cause of inequities (e.g. Ramírez & Pacheco, 2016; Barrance & Elwood, 2018). In this case, it is obvious that the responsibility for success is placed either on the individual learner or the social learning support. The next section therefore focuses on describing the factors that may influence educational equity in the Malaysian context.

Factors Influencing Educational Equity in the Malaysian Context

If educational equity is to be realised, the cause of inequity must be understood, and only then can adequate measures be undertaken to address equity. A point to note is that although this section briefly describes the factors influencing educational equity in general, it is also valid and applicable for science education in Malaysia. Based on the literature reviewed, three common and most significant factors influencing educational equity are described.

Allocation of Funds

There are three dominant views on the importance of educational resources to significantly improve students' achievement and address equity. The first view states that educational resources are irrelevant as it does not influence educational attainment. The core argument for the perspective on educational resources stemmed from the "Coleman Report", where Coleman et al. (1966) claimed that students' achievement is greatly influenced by their family background and that educational resources have little influence on it. Similarly, based on 300 empirical studies, Hanushek and Rivkin (2006) asserted that there is no strong or consistent relation between educational resources and student achievement. Meanwhile, the second view argues that educational funds do matter in improving educational outcomes. Challenging the former simplistic perspective on the intertwined relation between educational resources and student achievement, Hedges et al. (1994) conducted a meta-analysis on Hanushek's studies. Surprisingly, they found a statistically significant correlation between educational resources and student achievement. Hence, they suggested that large expenditure on education may positively increase student achievement. Supporting Hedges et al.'s (1994) argument, Verstegen and King (1998), who comprehensively and critically reviewed 35 years of research studies on educational resources and student achievement, confidently asserted that the amount allocated to schools can make a difference on students' achievement.

This situation is evident in Malaysia. With large spending on education that approached 5% of gross domestic product (GDP), it is clear that Malaysia greatly values the education of the future generation. The significant increase on the enrolment of Malaysian students to primary and secondary education is most likely due to Malaysia's high expenditure on education (Nasir et al., 2006). Although Malaysia should be enormously proud of its high enrolment rates, the achievement gap between the rich and poor, urban and rural still prevails. This indicates that higher levels of spending on education are not necessarily correlated with better outcomes. For instance, with less expenditure per pupil, Thailand's student achievements are better compared to Malaysia's own (Thien et al., 2015). This suggests that while a certain threshold of spending is required, it is more important that money is put directly towards the right factors in order to ensure success.

The third view strongly states it is not the quantity of educational resources that matters, but how the educational resources are distributed that is most important. In this perspective, Faubert and Blacklock (2012), and BenDavid-Hadar (2018), argued that educational resources may not matter for students from advantaged backgrounds; however, it matters the most for students from disadvantaged backgrounds. Hence, in order to embrace educational equity, BenDavid-Hadar (2018) suggested that allocation of educational resources should be carried out according to students' and schools' needs.

Malaysia has put in the effort to ensure that the students who need the most help are prioritised in getting support. Due to this, Malaysia has been giving more attention and focus on segments of society such as the Orang Asli and indigenous students, at-risk students and students with special needs by offering various educational pathways such as Islamic schools, art schools, and sport schools, and not forgetting, striving to have more than 30% of all Special Education Needs Students in the Inclusive Education Programme.

However, the author argues that it is timely for Malaysia to adopt or even devise its own targeted needs-based funding formula to ensure the allocation of educational resources is effective and is focused on students in need.

Access to Quality Teachers

The discussion now turns to highlight the important link between teachers and student outcome and how teacher qualifications relate to educational equity. Qualified teachers are commonly referred to as teachers who have the appropriate teaching certifications and experiences. It has long been recognised that teacher quality is a factor of utmost importance that contributes to school improvement and student achievement (Hanushek, 2002). Many researchers have suggested that students who benefit the most from having qualified and effective teachers are those in disadvantaged schools. Unfortunately, previous research on teacher employment (Simon & Johnson, 2015) indicates that highly trained and experienced teachers tend to be assigned to less challenging and easier-to-serve schools. This issue was raised again in a report by the Organisation for Economic Co-operation and Development (OECD, 2018) entitled Effective Teacher Policies: Insights from PISA - which found that in many developing countries, disadvantaged schools have less qualified teachers because of ineffective policies and unattractive incentives in retaining these high-quality teachers at the disadvantaged schools.

Similarly, Malaysia also faces the same problem. In Malaysia's highly centralised administration of education, its administrative structure is hierarchically categorised into four distinctive levels. They are federal, state, district and school. Although the Ministry of Education (MOE) Malaysia is responsible for allocating trained teachers to the states, it is up to the District Education Office (DEO) to decide to which school a teacher is posted. Therefore, in encouraging teachers to teach at low socio-economic and rural schools, the MOE contemplates payment of extra allowance to teachers who are willing to teach in rural schools. Unfortunately, evidenced by a high turnover rate of Malaysian teachers at rural schools, this strategy is far less than effective (Singh, 2014).

Exploring the aforementioned issue further, a great majority of teachers posted to rural schools highlighted the plight of the schools in terms of basic facilities such as clean water, electricity supplies, living amenities and poor teacher's quarters (Singh, 2014). As a result, most of the teachers were demoralised and wanted to be posted out despite incentives given by the government. Consequently, the constant change of teachers will less likely result in rural schools attracting and retaining high-quality teachers to serve and thus would affect the equity of education that the government has planned.

Curriculum, Teaching, and Assessment

As suggested by many researchers, curriculum and teaching are two major factors that have the potential to address equity in education (Westbrook et al., 2013). Therefore, curriculum and teaching should be seriously seen as key levers in improving students' academic achievement. With regard to equity, these countries are more likely to expose all students to high levels of education standard and expect all students to excel academically regardless of their background and starting points (Banerjee, 2016).

In the Malaysian context, comprehensive understanding of educational equity issues related to curriculum should be seen through the lens of Malaysia's ethnic composition (Heng & Tan, 2006). It is well known that one of Malaysia's education goals is to produce a competent and skilful workforce while integrating unity among the different ethnics in Malaysia. Nevertheless, due to strong discontent and dissent from the minorities, vernacular schools were established to complement its mainstream counterpart. However, instead of complementing the mainstream schools, these schools are gradually competing with their counterparts and asserting their own ideology in the national curriculum. As a result, the students undergo different learning experiences based on the type of school (Gill et al., 2013).

Additionally, private Islamic schools or *sekolah agama pondok* that are registered or those unregistered under the Ministry of Education Malaysia also widen education equity. Some of these Islamic schools separate academic subjects and focus too much on Islamic lessons to the extent of excluding national exams which causes their students to be left behind in the working field as they lack the required certificates (Islam et al., 2019).

Furthermore, issues on educational equity become even more critical with the introduction of high-performing schools such as Mara Junior Science College (MRSM) that adopt and innovate high-quality curricula and programmes, for example, through the inclusion of Schoolwide Enrichment Model (SEM), which indirectly widen the gap between students. Schoolwide Enrichment Model (SEM) that adds value to existing curricula are most welcomed and encouraged; however, these enrichment curricula should be offered and opened to every student rather than selected students (Kotok, 2017).

Although such is the situation, the Malaysian government should be praised for its various efforts to address the issues of equity among the minorities such as the Orang Asli and Penan communities through *Kurikulum Bersepadu Orang Asli/Penan*. However, this initiative is causing an educational deficit since it is necessary for the minority students following these specialised curriculums to sit for the standardised national examinations. Thus, it is impossible to achieve education equity when the curriculum, teaching and learning process as well as assessment practices are incongruent (Looney & Klenowski, 2008).

In discussing the assessment aspect, Malaysia is a nation that adopts a high-stake standardised examination system. This situation is widely known to be one of the main reasons why teachers are reluctant to venture beyond their comfort zone or apply innovative teaching or student-centred and learning-centred pedagogical approaches (Sidhu et al., 2011). The reason for this is that both of these approaches are time-consuming in terms of preparation and implementation. Teaching approach that fails to fulfil students' perspective, profiles, interest and ability has the potential to further widen educational inequity (e.g. Barrance & Elwood, 2018). Malaysia is currently moving away gradually from the high-stakes examination system to a system which will allow room for teachers to be innovative and creative in their teaching sessions.

After discussing the three major factors which may significantly influence educational equity in the Malaysian context, this chapter progresses to propose strategies that can be used to address educational equity.

How to Achieve Educational Equity through Culturally Responsive Science Pedagogy?

Although combating the aforementioned hindering factors is crucial, this chapter proposes that equitable education is also achievable at minimal cost if curriculum differentiation (CD) is upheld through the implementation of CRSP. This is because CD offers the best framework in addressing the needs of individual learners while CRSP provides clear guidelines on the science educational practices that have been proven effective in supporting students who have been historically unsuccessful in mainstream classrooms due to their cultural differences.

A point to note is that in recognising the importance of establishing equitable education, Malaysian teachers have been empowered by the Ministry of Education to meticulously adapt or modify the "formal or prescribed curriculum" to fit their students' learning needs. This suggests that although Malaysian teachers are working within the context of ministerial curriculum guidelines, they are granted a degree of flexibility to adapt and modify the curriculum, teaching approaches as well as the learning resources to ensure that all students in their classes learn to their potential (Coe et al., 2014). Some of the current educational changes introduced by the government are the integration of information and communications technology (ICT) in classroom teaching (Simin & Mohammed Sani, 2015), the teaching of higher order thinking skills, the implementation of classroom-based assessment (Hasnida & Ghazali, 2016) and the introduction of inclusive education programme (Nasir & Efendi, 2017), to name a few. What these newly introduced educational approaches and initiatives share in common are the key principles of CD where no one will be left behind if appropriate response and educational intervention are able to meet learners' diversity and needs (Jelas, 2010).

From the very beginning, with its root in brain research and multiple intelligence theory (Gardner, 2020), CD highly recognises individual students' differences and its influence on students' approach to learning. Differentiation is a widely known educational approach that not only acknowledges student differences in terms of their prior background knowledge, learning readiness, linguistic variety and language proficiency, types of learning style and interest but also makes an effort towards (a) modifying and adapting curriculum content, (b) designing responsive instructional approaches and (c) practising assessment that meet particular students' needs (Tomlinson, 1999; Hodge, 2020)

In Tomlinson's (1999) conceptualisation of CD, differentiated curriculum content refers to providing a variety of additional and supporting learning materials to ensure all students are able to achieve the predetermined learning objectives.

However, this proves to be a challenge for teachers particularly in the Malaysian context as teachers are responsible for aligning their teaching to cover the curriculum content to prepare students for the high-stakes standardised examinations (Asma Iffah et al., 2013). In contrast, differentiated assessment practices focus on ongoing and formative assessment which enables teachers to offer a buffet of assessment choices and scaffolds within the many needs, interests and abilities of students in an academically diverse classroom. These practices are sometimes seen as unfair (Rubie-Davies et al., 2010) and create discrimination to some students as teachers generally tend to have lower expectations with low-ability students and commonly assign simpler tasks to them (Rubie-Davies et al., 2010). Moreover, for teachers, preparing, administering and assessing individualised assessment approaches are time consuming (Hasnida & Ghazali, 2016). Finally, Tomlinson's (1999) differentiated instructional approaches demand teachers to design various teaching approaches and create an active learning atmosphere which aims to maximise students' learning regardless of their background. Realistically, the latter approaches to CD which focus on designing appropriate instructional strategies to meet students' diverse learning needs hold the promise for practical and pragmatic usage within the classroom (Singh, 2014).

Nonetheless, despite extensive research that has been carried out to popularise CD and in particular differentiated instructional approaches, CD scholars fail to provide teachers with specific guidelines on how to adapt and modify their methods of teaching according to the different ability levels and needs of the students. Additionally, coupled with the vague definition of what curriculum differentiation means, many misconceptions on CD arise and among the common CD misconceptions are: (1) CD focuses on catering to the individual learning needs of students; although this will be better for students, we would argue that it ignores the realities of the actual classroom size, especially the classrooms in Malaysia which usually consist of up to 40 students; and (2) CD provides different learning objectives, different learning expectations and completely different learning activities for students with different needs. This is a dumbing-down strategy where teachers have low expectations towards low-ability students and assign simpler tasks to these students, and indeed this practice is a 360-degree contradiction of the concept of equity. It is these identified gaps that lead to the impetus to suggest for the implementation of culturally responsive pedagogy to serve as a bridge in offering improved and effective teaching practices for all students in the differentiated classroom.

As we all know, our preferences, interests and abilities are significantly influenced by our cultural background which originates not only from ethnicity, religion, gender and race but also from the family, local communities and the country (Fong et al., 2016). Therefore, having an awareness of how culture influences one's ability to learn is important for teachers in ideally designing meaningful and effective learning experiences for all students.

However, this is not always the case in schools. Despite the different learning needs, abilities, interest and most importantly, rich background diversity, students are expected to learn the same curriculum content in one particular or identical way (Banerjee, 2016). As a result, students are barely able to relate the relevance of classroom learning to their out-of-school living. The importance of ensuring the connection between in-school learning and out-of-school living has been cited by many researchers as the key ingredient of meaningful learning (James & Williams, 2017) which ultimately improves students' academic performance. This way of teaching and learning that stresses on students' background is referred as CRSP, which is one of the differentiated instructional strategies under the larger concept of CD.

Conclusion

Based on the aforementioned discussion, the author suggests two pragmatic approaches that can be used to improve equity in education which are educational resources and practices. First is related to educational resources. The resources would encompass (a) education funding for students with greater needs, (b) complete and latest basic school facilities and infrastructure to support teaching and learning activities, (c) improving and enriching the quality of teachers through consistent professional training and (d) providing access to high-quality education for all students. Second is related to practices and comprises (a) differentiating instruction through CRSP to meet students' background, (b) strengthening the links between school and home to support disadvantaged parents to help their children learn, (c) responding attentively to students' background for successful inclusion of vulnerable minorities within mainstream education and (d) providing systematic help or safety net to those who are at risk of falling behind at schools. In conclusion, despite various ways to address equity in education, the author proposes that CRSP is the most relevant option because it is in the teachers' control and power. Furthermore, the financial cost to execute this approach is the lowest compared to other options. Most importantly, CRSP has been proven to be able to provide meaningful learning experiences for students, thus making them successful in learning which indirectly reduces the equity gap.

References

Ainscow, M. (2020). Inclusion and equity in education: Making sense of global challenges. *Prospects: Comparative Journal of Curriculum, Learning and Assessment, 49*, 123–134.

Asma Iffah, Z. N., Abd Samad, A., & Omar, Z. (2013). Pressure to improve scores in Standardized English Examinations and their effects on classroom practices. *International Journal of English Language Education, 2*(1), 45–56. https://doi.org/10.5296/ijele.v2i1.4524

Avendano, L., Jessica, R., Sarah, K., & Kamal, H. (2019). Bringing equity to underserved communities through STEM education: Implications for leadership development. *Journal of Educational Administration and History, 51*(1), 66–82.

Banerjee, P. A. (2016). A systematic review of factors linked to poor academic performance of disadvantaged students in science and math in schools. *Cogent Education, 3*, Article 1178441. https://doi.org/10.1080/2331186X.2016.1178441

Barrance, R., & Elwood, J. (2018). Young people's views on choice and fairness through their experiences of curriculum as examination specifications at GCSE. *Oxford Review of Education, 44*(1),19–36.

BenDavid-Hadar, I. (2018). Funding education: Developing a method of allocation for improvement. *International Journal of Educational Management, 32*(1), 2–26.

Coe, R., Aloisi, C., Higgins, S., & Major, L. E. (2014). *What makes great teaching? Review of the underpinning research.* Sutton Trust. www.suttontrust.com/wp-content/uploads/2014/10/What-Makes-GreatTeaching-REPORT.pdf

Coleman, J. S., Campbell, E. Q., Hobson, C. J., McPartland, J., Mood, A. M., Weinfeld, F. D., & York, R. L. (1966). *Equality of educational opportunity.* U.S. Government Printing Office.

Department of Statistics Malaysia. (2018). Children Statistics Publication, Malaysia. https://www.dosm.gov.my/v1/index.php?r=column/pdfPrev&id=RWsxR3RwR VhDRlJkK1BLalgrMGRlQT09#:~:text=Children%20by%20age%20group%20and %20sex&text=%2C%20the%20composition%20of%20children%20under,than%20 female%20(4.55%20million). Accessed May 20, 2021.

Faubert, B., & Blacklock. (2012). *Review of evaluation studies on reducing failure in schools and improving equity* [Project analytical paper]. OECD. www.oecd.edu/equity

Field, S., Kuczera, M., Pont, B., & Organisation for Economic Co-operation and Development. (2007). *No more failures: Ten steps to equity in education.* OECD.

Fong, E. H., Catagnus, R. M., Brodhead, M. T., Quigley, S., & Field, S. (2016). Developing the cultural awareness skills of behavior analysts. *Behavior Analysis in Practice, 9*(1), 84–94.

Gardner, H. (2020). Neuromyths: A critical consideration. *Mind Brain Education, 14*, 2–4. https://doi.org/10.1111/mbe.12229

Gill, S. K., Keong, Y. C., Beng, C. O. S., & Yan, H. (2013). Impact of Chinese vernacular medium of instruction on unity in multi-ethnic Malaysia. *Pertanika Journal of Social Science and Humanities, 21*(3), 1039–1064.

Goolamally, N., & Ahmad, J. (2010). *Boys do poorly in schools: The Malaysian story.* Unpublished Manuscript. https://www.researchgate.net/publication/262689050_ BOYS_DO_POORLY_IN_SCHOOLS_THE_MALAYSIAN_STORY. Accessed May 20, 2021.

Hanushek, E. A. (2002). Evidence, politics, and the class size debate. *The Class Size Debate*, 37–65.

Hanushek, E. A., & Rivkin, S. G. (2006). Teacher quality. *Handbook of the Economics of Education, 2*, 1051–1078.

Hasnida, N., & Ghazali, C. M. (2016). The implementation of school-based assessment system in Malaysia: A study of teacher perceptions. *Geografia: Journal of Society and Space, 12*(9), 104–117.

Hedges, L. V., Laine, R. D., & Greenwald, R. (1994). An exchange: Part I: Does money matter? A meta-analysis of studies of the effects of differential school inputs on student outcomes. *Educational Researcher, 23*(3), 5–14.

Heng, C. S., & Tan, H. (2006). English for mathematics and science: Current Malaysian language-in-education policies and practices. *Language and Education, 20*(4), 306–321.

Hodge, E. M. (2020). Conceptions of equity in common core policy messages in a metropolitan district. *The Educational Forum, 85*(1), 89–106. https://doi.org/ 10.1080/00131725.2020.1772427

Howard, T. C. (2013). How does it feel to be a problem? Black male students, schools, and learning in enhancing the knowledge base to disrupt deficit frameworks. *Review of Research in Education, 37*(1), 54–86. https://doi.org/10.1787/9789264214033-en

Islam, R., Haidoub, I. M., & Tarique, K. M. (2019). Enhancing quality of education: A case study on an international Islamic school. *Asian Academy of Management Journal, 24,* 141–156. https://doi.org/10.21315/AAMJ2019.24.S1.10

James, J. K., & Williams, T. (2017). School-based experiential outdoor education. *Journal of Experiential Education, 40*(1), 58–71. https://doi.org/10.1177/1053825916676190

Jelas, Z. M. (2010). Learner diversity and inclusive education: A new paradigm for teacher education in Malaysia. *Procedia – Social and Behavioral Sciences, 7*(C), 201–204. https://doi.org/10.1016/j.sbspro.2010.10.028

Kotok, S. (2017). Unfulfilled potential: High-achieving minority students and the high school achievement gap in math. *The High School Journal, 100,* 183–202. https://doi.org/10.1353/hsj.2017.0007

Looney, A., & Klenowski, V. (2008). Curriculum and assessment for the knowledge society: Interrogating experiences in the Republic of Ireland and Queensland, Australia. *Curriculum Journal, 19*(3), 177–192. https://doi.org/10.1080/09585170802357496

Nasir, M. N. A., & Efendi, A. N. A. E. (2017). Special education for children with disabilities in Malaysia: Progress and obstacles Muhamad Nadhir Abdul Nasir. *Geografia-Malaysian Journal of Society and Space, 12*(10).

Nasir, N. I. S., Rosebery, A. S., Warren, B., & Lee, C. D. (2006). *Learning as a cultural process: Achieving equity through diversity* (pp. 489–504). The Cambridge Handbook of the Learning Sciences.

OECD. (2018). *Effective teacher policies: Insights from PISA.* OECD Publishing. https://doi.org/10.1787/9789264301603-en.

Osman, K., & Marimuthu, N. (2010). Setting new learning targets for the 21st century science education in Malaysia. *Procedia-Social and Behavioral Sciences, 2*(2), 3737–3741. https://doi.org/10.1016/j.sbspro.2010.03.581

Ramírez, O. M., & Pacheco, S. D. L. T. (2016). Equity and quality in education from a student perspective: Semantic constructs in Chilean undergrads. *Journal of Social Science for Policy Implications, 4*(2), 7–21. https://doi.org/10.15640/jsspi.v4n2a2

Rubie-Davies, C. M., Peterson, E., Irving, E., Widdowson, D., & Dixon, R. (2010). Expectations of achievement. *Research in Education, 83*(1), 36–53. https://doi.org/10.7227/RIE.83.4

Sidhu, G. K., Fook, C. Y., & Mohamad, A. (2011). Teachers' knowledge and understanding of the Malaysian school-based oral English assessment. *Malaysian Journal of Learning and Instruction, 8,* 93–115.

Simin, G., & Sani, Ibrahim Mohammed. (2015). Effectiveness of ICT integration in Malaysian Schools: A quantitative analysis. *International Research Journal for Quality in Education, 2*(8), 1–12.

Simon, N. S., & Johnson, S. M. (2015). Teacher turnover in high-poverty schools: What we know and can do. *Teachers College Record, 117*(3), 1–36.

Singh, H. (2014). Differentiating classroom instruction to cater to learners of different styles. *Indian Journal of Applied Research, 3*(12), 58–60. https://doi.10.15373/22501991/December2014/25

Thien, L. M., Darmawan, I. G. N., & Ong, M. Y. (2015). Affective characteristics and mathematics performance in Indonesia, Malaysia, and Thailand: What can PISA 2012 data tell us? *Large-Scale Assess Education, 3*(3). https://doi.org/10.1186/s40536-015-0013-z

Tomlinson, C. (1999). *The differentiated classroom: Responding to the needs of all learners.* Association for Supervision and Curriculum Development.

Verstegen, D. A., & King, R. A. (1998). The relationship between school spending and student achievement: A review and analysis of 35 years of production function research. *Journal of Education Finance, 24*(2), 243–262.

Westbrook, J., Durrani, N., Brown, R., Orr, D., Pryor, J., Boddy, J., & Salvi, F. (2013). *Pedagogy, curriculum, teaching practices and teacher education in developing countries.* Final Report. Education Rigorous Literature Review. Department for International Development.

4 Culturally Responsive Pedagogy

A Framework for Science Teaching in Asian Contexts

Edy Hafizan Mohd Shahali, Lilia Halim and Mohd Ali Samsudin

Introduction

Many students experience a disconnect between what is taught in the science classroom and their cultural backgrounds. This disconnect may explain why some students are less likely to engage and excel in science. Research suggests that challenges in science learning increase for students who do not bring the same views and ways of knowing science from their culture as are taught in schools (Atwater et al., 2010; Bryan & Atwater, 2002; Bianchini et al., 2003).

Wilson (1981, p. 29) highlighted that "for science education to be relevant to a specific context, it must take much more explicit account of the cultural context of the society which provides its setting, and whose needs it exists to serve". Odora-Hoppers (2001) argues that the nature of science education in most Asian countries is still dominated by logical positivism with the imposition of Western values and epistemologies. It can also be argued that the status of science education in many Asian countries has remained hegemonic, absolutist and universalistic. Thus, one could argue that if science education in Asian countries lacks relevance to local cultures, then students would view learning science as being dominated by a collection of facts.

In addition, Arsad et al. (2020) in a systematic literature review on culturally relevant science teaching (CRST) found that "Asian countries lack of CRST research that applies socio-culture accountability – which refers to the use of the science content knowledge and understanding, skills and attitude that empower students socially, intellectually and politically" (p. 4). Thus, it could be argued that acquisition of scientific knowledge, thinking and practices appear to have little practical utility in the eyes of the students. This scenario has resulted in the alienation of the recipients of science (students) from their environment There is a need, therefore, to develop a culture and context sensitiveness within science education in Asian environments.

What Is Culturally Responsive Pedagogy?

Culturally responsive pedagogy (CRP) relates to the kind of education that establishes the usefulness of what is taught in real-life examples, relating materials

DOI: 10.4324/9781003168706-5

to everyday applications, and giving local examples (Kember & McNaught, 2007). CRP uses "the cultural characteristics, experiences, and perspectives of ethnically diverse students as conduits for teaching them more effectively" (Gay, 2002, p. 106). Characteristics of culturally responsive pedagogy practices focus on meeting the needs of culturally, ethnically and racially diverse students. It includes, but is not limited to, building rapport with students, building bridges between school and home, creating a positive culture, using multicultural literature to engage students and differentiating instruction (Gay, 2010). In all cases, CRP is based on the idea that cultural (and linguistic) practices of underrepresented students are assets rather than deficits or barriers to the learning process. It is well established in many studies that CRP does not only help students develop the necessary learning skills, but teachers that use this approach will also be able to create social relationships with their students to promote a safe and nurturing environment that is inclusive for learning.

Theoretically, CRP shares the features of constructivist, student-centred and authentic learning practice (e.g. by connecting course content with students' prior knowledge) (Richardson, 2003; Yilmaz, 2008). CRP is also deeply embedded in students' everyday lives, their communities and cultural funds of knowledge (FoK). This would enable students to see how the subjects fit together and eventually contribute to reducing achievement gaps and promoting positive ethnic-racial identities for culturally diverse and underrepresented students in science (Dickson et al., 2015; Sleeter, 2012). Education must aim for equity and make deliberate attempts to meet students' cultural requirements in an ethnically varied setting (Paris, 2012). However, the current educational system in many Asian countries places too much focus on Eurocentrism, which is reflected in the curriculum and instruction, that does not foster culturally responsive teaching (Ladson-Billings, 2014). As demographics in the Asian countries are diverse, an educational reform that demands for culturally responsive pedagogy is required (Lyons et al., 2016).

Culturally Responsive Pedagogy and Science Teaching

Scholars recognise the need to increase relevance in science education among students in general and underrepresented students in science in particular. Various efforts (theoretically or pedagogically) are being proposed to meet the needs and interests of the students, of student careers and of society and the culture. The authors hypothesised these four concepts (i) culturally responsive pedagogy, (ii) contextual teaching and learning, (iii) FoK and (iv) students' prior knowledge to serve as the basis of the CRP framework in science teaching – from now on is known alternatively as Culturally responsive science pedagogy (CRSP) (Figure 4.1).

This framework, as a whole, suggests that students' prior knowledge and experiences that they bring into science classroom learning are related to the FoK the students acquire from their cultural background and interaction with the community. The concept of contextual teaching and learning suggests the need to acknowledge that contextual learning should focus not only on career

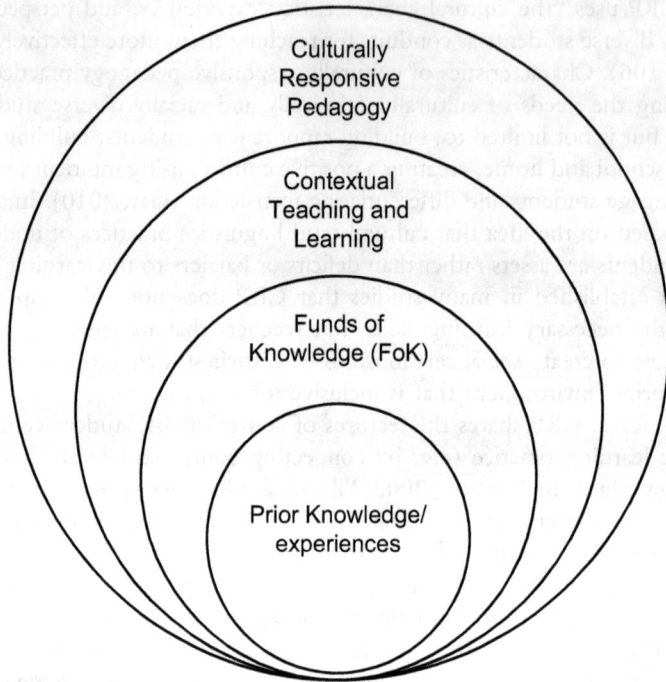

Figure 4.1 Culturally responsive science pedagogy framework

development but also on encouraging students to be critical of their surroundings and community and help to improve their community for the better.

Finally, considering the aforementioned concepts will lead to thinking about science teaching as a pedagogical approach that is culturally responsive. In other words, this framework advocates that science teachers consider students' culture in every aspect of their instructional approach or strategies and not in ad-hoc fashion to their teaching.

These four elements depicted in Figure 4.1 can further be categorised into two-layered perspectives: (a) cognitive and (b) socio-political consciousness.

Cognitive Science perspective

From the cognitive science perspective, teachers need to identify points of contact where scientific practices are continuous with students' everyday knowledge and build on such continuities to promote student learning (Warren et al., 2001). The authors propose these two elements as the points of contact that will bridge students' learning of scientific concepts seamlessly with their diverse backgrounds, knowledge and experiences.

Element 1: Students' Prior Knowledge

Students' prior knowledge and experiences are crucial components in acquiring new information. Students bring different levels of knowledge into the science classroom; therefore, it is important to show relevance during the teaching and learning processes. Drawing upon students' prior knowledge or experiences in the content helps teachers link the science learning goals with what students already know (Bianchini & Brenner, 2010). This prior knowledge is the accumulation of students' prior knowledge that comes from their cultural experiences. In addition, as highlighted by Arsad et al. (2020) "Cultural backgrounds do not limit to indigenous culture of a group but cultural backgrounds also refer to practices, daily routine or work experience that are contemporary as well as based on families' inner culture" (p. 2).

Element 2: Funds of Knowledge

FoK are collections of knowledge based on cultural practices that are a part of families' inner culture, work experience or daily activities. It is the knowledge and expertise that students have because of their roles in their families, communities and culture. Along the same line, scholars have defined FoK as "historically accumulated and culturally developed bodies of knowledge and skills essential for household or individual functioning and well-being" (Gonzalez et al., 2005, p. 72). A "funds of knowledge" framework looks at a student's experiences as assets rather than deficits. However, adoption of this new focus is still weak in the science classroom where implementing CRSP often seems to encourage the promotion of misconceptions of the content. Thus, CRSP requires a firm understanding of the content on the part of the teachers. In addition, teachers are to understand their students' lives and communities (i.e. develop teachers' cultural competencies), thus being able to create the bridge between Western science and the students' FoK. As argued by Byrd (2016), the philosophy of CRSP does not aim to provide learning concessions for students of culturally diverse groups. Instead, CRSP expects students, who are often marginalised in science learning, to gain high academic competence.

Socio-political Consciousness

The literature on the socio-political consciousness perspective in CRSP (Rodriguez & Berryman, 2002) suggests that teachers need to jointly engage in a critical analysis of the purposes of schooling and of science with the students before the teachers engage the students in learning science. The authors also suggest that adopting the socio-political consciousness perspective is to encourage students to improve their community and surroundings (which may or not be resulted from inequitable distribution of power and social class) by drawing on the scientific knowledge, thinking and practices. In this perspective, the two related elements are (1) contextual teaching and learning and (2) culturally responsive pedagogy.

Element 3: Contextual Teaching and Learning

Contextual teaching and learning involve making learning meaningful to the students by connecting it to the real world. It draws upon students' interests, experiences, diverse skills and cultures and integrates these into what and how students learn and how they are assessed. Meaningful contextual teaching situates learning and learning activities in real-life and vocational contexts to which students can relate, incorporating not only content, that is, the "what," of learning but also the reasons why learning is important. Interdisciplinary activities across content areas, classrooms and grade levels, or among students, classrooms and communities are some examples of contextual teaching and learning in the science classroom.

Element 4: Culturally Responsive Pedagogy

The diversity of students can lead to challenges for teachers, especially those not familiar with CRP (Lew & Nelson, 2016). CRP is the lens through which teachers approach their work. It focuses on the academic and personal success of students by ensuring that students engage in academically rigorous curriculum and learning. Through CRP, students are also empowered to transform society through critical consciousness among the students in relation to scientific issues and problem solving in the community (e.g. Epstein et al., 2011; Morrell & Duncan-Andrade, 2002).

Challenges and Strategies of Effective CRP in Science Teaching

Even though CRP in science teaching or CRSP can help to lessen the alienation of science from culturally diverse students in terms of ethnicity, gender, socio-economic status and linguistically, various challenges have been reported. Among the challenges reported are what to include and exclude in science education of the specific setting in order to be responsive to the context and culture especially for the learners (Aikenhead, 2006). These challenges and ongoing strategies in addressing the challenges are summarised further:

Curriculum

Curriculum relevance should reflect a part of education that attempts to keep school children in touch with the common people, with their parents and their communities. The most important point to recognise is that a one-size-fits-all curriculum and instructional approach will surely fail to meet the learning needs of the students in any classroom. It is therefore imperative that curriculum developers clearly develop science education policies that reflect learner-centred frameworks of pedagogy that are responsive to learners' needs and experiences (Gay, 2000; Osborne & Collins, 2001; Sjoberg & Schreiner, 2003).

Books and Learning Materials

Publication (books and learning materials) of the knowledge and experiences needs to be contextually and culturally responsive and designed to meet the needs of the specific community. Teaching and learning materials such as in Malaysia and Indonesia are beginning to recognise the need to relate the materials to their own culture, community and local surroundings. This has been practised by Japanese science education, whereby investigation of science concepts and applying scientific process skills are conducted and related to their local and national needs, issues and problems. Thus, providing texts that are not connected to experiences of the non-Western communities should be an ongoing effort and agenda.

Students' Readiness

Previous studies revealed that most students in Asian contexts are generally attentive to their science lessons and spend a lot of time studying and looking for the right answers; yet real understanding of the subject has appeared elusive to many. This situation seems to have emanated from the unresponsiveness of science education to the cultural context of learners (Barton et al., 2003); and science subjects that often fail to link the subjects with prior knowledge that students have acquired from their different indigenous and prior knowledge.

As a result, students often spend hours memorising de-contextualised lists, formulae, and teacher-directed procedures and activities which make little sense but are seen by students and teachers as essential to the education process. The science education they receive, therefore, will likely push the students away from science (Singar & Zainuddin, 2017) and become uncritical consumers of the ideas (Lewin, 1993).

Assessment

Studies argued that most national examination boards in schools assume an ideal situation and set the same national exam without any regard for the local conditions of learning in each school (Anamuah-Mensah et al., 2004). However, in the case of Malaysia, the need to be inclusive in assessment procedure is beginning to emerge. The national examination board in realising inclusivity in assessment procedures allow indigenous terms to be used in providing solutions, and it is included in the marking scheme. For example, in explaining a scientific phenomenon or process, a local indigenous term is used. The term is considered acceptable for grading purposes as long as the explanation provided in the answer is scientifically correct.

Science Pedagogy

Science knowledge is always presented as an established, precise and non-challenged course of study which leaves no room for learners in these contexts to question its viability (Snively & Corsiglia, 2001). However, research has demonstrated that learners would be capable of excelling well in science if they are given culturally appropriate instruction. Culturally appropriate instruction would help students who are grounded in their local traditional knowledge to clarify the science concepts. Unfortunately, many teachers do not recognise or value their students' cultural knowledge and often neglect to affirm their students' traditional teachings which are nurtured in the home and the local community. As a result, these students lose their desire to learn science, avoid professions in science and question their own cultural identity.

Among the CRSP practices in science classrooms are:

- Students use scientific principles to analyse and address social injustices such as local water pollution (Dimick, 2012).
- Students make connections between traditional cultural practices, such as arrow making and throwing, and science content, such as accelerated motion (Grimberg & Gummer, 2013).
- Students generate their own lines of scientific inquiry which are combined with authentic models of scientific inquiry (Buxton, 2006).
- Students explicitly identify how their linguistic and cultural experiences and values relate to those of science via instructional congruence (Lee & Fradd, 1998).

Language of Instruction in Science Classrooms

Most science educators agree that the greatest barrier to learning science in most non-Western countries is language (McKinley& Keegan, 2008). Using English as a language for science instruction has resulted in the majority of learners failing to comprehend what is written or taught, causing them to resort to memorising as a way of learning the subject. It is an undeniable fact that many indigenous languages have been endangered because of the use of English as an official language of instruction in most schools in non-Western countries since the nineteenth century (McKinley & Keegan, 2008). Consequently, this resulted in many students in developing countries having to develop the feeling that school science is like a foreign culture to them (Maddock, 1981). Their feeling stems from the fundamental cultural clashes between students' life-worlds and the world of Western science as a culture of its own as noted by Keane (2008).

In the context of Asian countries such as Malaysia, the medium of instruction in science and mathematics is in Malay. In 2003, the government decided to switch the medium of instruction from Malay to English as the medium of instruction in science and mathematics because English is seen as the language of scientific and technological knowledge (Halim & Meerah, 2016). Major studies (Ong & Tan, 2008; Hudson, 2009; Na & Mostafa, 2009) revealed that the policy of teaching science and mathematics in English has further widened the achievement gap between students in urban and rural schools in terms of achievement in science, with the rural students often lagging behind their urban counterpart. One of the many reasons for the decline is the difficulty of teachers to deliver the subject matter in the form that can be understood by learners in a language which is either a second or third language to the majority of Malaysian learners.

In addition, Malaysia and Indonesia are multi-ethnic countries. In the case of Malaysia, students speak in their mother tongue at home which would be Malay, varieties of Chinese' dialects or Tamil (these are the three main ethnic groups in Malaysia) and numerous languages of the minority groups in Sabah and Sarawak in particular (Halim & Meerah, 2016). Malay as the medium of instruction of the dominant group in Malaysia could also affect the learning of science of the other ethnic groups. Similarly, for Indonesia, where the medium of instruction is Bahasa Indonesia, which may be foreign to other ethnic groups in Indonesia. In other words, the dominant local language of the country could also contribute and lead to inequitable opportunities to culturally and economically diverse ethnic groups to learn and engage in science education effectively.

The aforementioned challenges imply that curriculum, assessment, pedagogy and teaching materials must be continuously improved to influence students' interests, attitudes and engagement with science learning as studies have revealed that there is a correlation between the way science is taught to students with the images of science the students depict. The authors argue that if teachers and schools are oblivious of cultural differences, it can lead to student and school failure (Nieto & Bode, 2008).

Research is fairly consistent on the qualities of a culturally responsive teacher (Byrd, 2016). Teachers who are culturally responsive will normally assess their own ability to communicate effectively with their students to increase the students' academic performance. Once teachers have become culturally aware (not necessarily competent or proficient), they will be able to effectively communicate with their students and help to increase science achievement within the subgroups (Paulk et al., 2014). Teachers that can "read" their students' actions for signs of learning will be able to provide equitable learning opportunities for all students, particularly those who have traditionally been marginalised in science classrooms.

There is a growing body of research providing guidance on how to effectively prepare culturally responsive science teachers, be it through teachers (i) critically reflecting on their own practices, (ii) actively participating as learners in culturally responsive science exemplars or (iii) being supported through the curriculum design process (Brown & Crippen, 2016). As suggested by Meerah et al. (2011), a professional development programme for primary science teachers, based on a collaborative action research concept, was able to improve their pedagogical skills namely in employing alternative teaching and learning approaches suitable for marginalised children. Teacher development in culturally responsive science teaching can also be supported through designing theoretically and contextually grounded induction and pre-service courses to support culturally responsive science teacher development (Brown et al., 2018). Through both types of courses, the science teachers will be involved in activities, discussions and reflections that raise awareness of the importance of attending to attitudes about culturally diverse students, as well as abilities to incorporate students' backgrounds into science instruction.

Conclusion

CRSP is a way of teaching that empowers students and incorporates their cultures, backgrounds and experiences into the school environment and activities of the science classroom. The proposed cultural pedagogy in science teaching framework is characterised by a number of different elements with two-layered perspectives. Both perspectives aim to (a) provide underrepresented students in science the opportunities to be successful academically in science and (b) encourage the students also to address related scientific problems in their community, thus being able to participate in decision making that is inclusive and equitable to their needs. The framework sets to encourage science teachers to reflect on their current practices and beliefs in science teaching that currently might be incomprehensive in meeting the needs (intellectually and socially) of students in their classrooms – who come from diverse backgrounds be it culturally, linguistically or economically. Each element in the framework depicts the interconnected complementary role that it can offer in offering an equitable science learning opportunity for students that are culturally diverse – with a view that CRSP is a way of thinking and doing, in relation to teachers' instructional practices, that eventually incorporate all the elements. This framework is relevant for both the Western and Asian contexts – as the framework explicitly links the elements and concepts related to science teaching within the realm of culture and learning.

References

Aikenhead, G. S. (2006). *Science education for everyday life: Evidence-based practice.* Teachers College Press.

Anamuah-Mensah, J., Mereku, D. K., & Ameyaw-Asabere, A. (2004). *Ghanaian junior secondary school students' achievement in mathematics and science: Results from Ghana's participation in the 2003 trends in international mathematics and science study.* Ministry of Education Youth and Sports.

Arsad, N. M., Nasri, N. M., Soh, T. M. T., Mahmud, S. N. D., Talib, M. A. A., & Halim, L. (2020, April). A systematic review on culturally relevant science teaching: Trends and insights. *AIP Conference Proceedings, 2215*(1), 040003. AIP Publishing LLC. https://doi.org/10.1063/5.0000530

Atwater, M., Freeman, T., Butler, M., & Draper-Morris, J. (2010). A case study of science teacher candidates understandings and actions related to the culturally responsive teaching of "other" students. *International Journal of Environmental & Science Education, 5,* 287–318.

Barton, A. C., Ermer, J., Burkett, T., & Osborne, M. (2003). *Teaching science for social justice.* Teachers College Press.

Bianchini, J. A., & Brenner, M. E. (2010). The role of induction in learning to teach toward equity: A study of beginning science and mathematics teachers. *Science Education, 94*(1), 164–195. https://doi.org/10.1002/sce.20353

Bianchini, J. A., Cavazos, L. M., & Rivas, M. (2003). At the intersection of contemporary descriptions of science and issues of equity and diversity: Student teachers' conceptions, rationales, and instructional practices. *Journal of Science Teacher Education*, *14*(4), 259–290. https://doi.org/10.1023/B:JSTE.0000009550.91975.76

Brown, J. C., & Crippen, K. J. (2016). Designing for culturally responsive science education through professional development. *International Journal of Science Education*, *38*(3), 470–492.

Brown, J. C., Ring-Whalen, E., Roehrig, G., & Ellis, J. (2018). Advancing culturally responsive science education in secondary classrooms through an induction course. *International Journal of Designs for Learning*, *9*(1), 14–33.

Bryan, L., & Atwater, M. (2002), Teacher beliefs and cultural models: A challenge for science teacher preparation programs. *Science Education*, *86*, 821–839.

Buxton, C. (2006). Creating contextually authentic science in a "low-performing" urban elementary school. *Journal of Research in Science Teaching*, *43*(7), 695–721. https://doi.org/10.1002/tea.20105

Byrd, C. M. (2016). Does culturally relevant teaching work? An examination from student perspectives. *SAGE Open*, *6*(3). https://doi.org/10.1177/2158244016660744

Dickson, G. L., Chun, H., & Fernandez, I. T. (2015). The development and initial validation of the student measure of culturally responsive teaching. *Assessment for Effective Intervention*, *41*, 141–154. https://doi.org/10.1177/1534508415604879

Dimick, A. S. (2012), Student empowerment in an environmental science classroom: Toward a framework for social justice science education. *Science Education*, *96*, 990–1012. https://doi.org/10.1002/sce.21035

Epstein, T., Mayorga, E., & Nelson, J. (2011). Teaching about race in an urban history class: The effects of culturally responsive teaching. *Journal of Social Studies Research*, *35*(1), 2.

Gay, G. (2000). *Culturally responsive teaching: Theory, research, and practice*. Teachers College Press.

Gay, G. (2002). Preparing for culturally responsive teaching. *Journal of Teacher Education*, *53*, 106–16.

Gay, G. (2010). *Culturally responsive teaching: Theory, research and practice*. Teachers College Press. Teachers College Columbia University.

González, N., Moll, L., & Amanti, C. (2005). *Funds of knowledge: Theorizing practices in households, communities, and classrooms*. Lawrence Erlbaum.

Grimberg, B., & Gummer, E. (2013). Teaching science from cultural points of intersection. *Journal of Research in Science Teaching*, *50*(1), 12–32. https://doi.org/10.1002/tea.21066

Halim, L., & Meerah, T. S. M. (2016). Science education research and practice in Malaysia. In M. H. Chiu (Ed.), *Science education research and practice in asia*. Springer. https://doi.org/10.1007/978-981-10-0847-4_5

Hudson, P. (2009). Learning to teach science using English as the medium of instruction. *Eurasia Journal of Mathematics, Science & Technology Education*, *5*(2), 165–170.

Keane, M. (2008). Science education and worldview. *Cultural Studies of Science Education*, *3*(3), 587–621.

Kember, D., & McNaught, C. (2007). *Enhancing university teaching: Lessons from research into award-winning teachers*. Routledge.

Ladson-Billings, G. (2014). Culturally relevant pedagogy 2.0: a.k.a. the remix. *Harvard Educational Review*, *84*(1), 74–84.

Lee, O., & Fradd, S. H. (1998). Science for all, including students from non-English-language backgrounds. *Educational Researcher, 27*(4), 12–21.

Lew, M. M., & Nelson, R. F. (2016). New teachers' challenges: How culturally responsive teaching, classroom management, & assessment literacy are intertwined. *Multicultural Education, 23*(3–4), 7.

Lewin, K. M. (1993). Education and development: The issues and the evidence. *Education Research. Serial No. 6.*

Lyons, R., Dsouza, N., & Quigley, C. (2016). Viewing equitable practices through the lens of intersecting identities. *Cultural Studies of Science Education, 11*(4), 941–951. https://doi.org/10.1007/s11422-015-9699-z

Maddock, M. N. (1981). Science education: An anthropological viewpoint. *Studies in Science Education, 8*(1), 1–26. https://doi.org/10.1080/03057268108559884

McKinley, E., & Keegan, P. J. (2008). Curriculum and language in Aotearoa New Zealand: From science to putaiao. *L1 Educational Studies in Language and Literature*, Special Issue.

Meerah, T. S. M., Halim, L., Rahman, S. A., Harun, H., & Abdullah, R. T. (2011). Teaching marginalized children primary science teachers' professional development through collaborative action research. *Cypriot Journal of Educational Sciences, 2*(1), 49–60.

Morrell, E., & Duncan-Andrade, J. M. (2002). Promoting academic literacy with urban youth through engaging hip-hop culture. *English Journal*, 88–92.

Na, C. L., & Mostafa, N. A. (2009). Teacher beliefs and the teaching of mathematics and science in English. *English Language Journal, 3*, 83–101.

Nieto, S., & Bode, P. (2008). *Affirming diversity: The socio-political context of multicultural education*. Pearson Education.

Odora-Hoppers, C. (2001). Indigenous knowledge systems and academic institutions in South Africa. *Perspectives in Education, 19*(1), 73–85.

Ong, S. L., & Tan, M. (2008). Mathematics and science in English: Teachers experience inside the classroom. *Jurnal Pendidik dan Pendidikan* [Education and Educator Journal], *23*, 141–150.

Osborne, J., & Collins, S. (2001). Pupils' views of the role and value of the science curriculum: A focus-group study. *International Journal of Science Education, 23*(5), 441–467.

Paris, D. (2012). Culturally sustaining pedagogy: A needed change in stance, terminology, and practice. *Educational Researcher, 41*, 93–97.

Paulk, S. M., Martinez, J., & Lambeth, D. T. (2014). Effects of culturally relevant teaching on seventh grade African American students. *MLET: The Journal of Middle Level Education in Texas, 1*(1), 3.

Richardson, V. (2003). Constructivist pedagogy. *Teachers College Record, 105*(9), 1623–1640.

Rodriguez, A. J., & Berryman, C. (2002). Using socio-transformative constructivism to teach for understanding in diverse classrooms: A beginning teacher's journey. *American Educational Research Journal, 39*(4), 1017–1045.

Singar, S. N., & Zainuddin, A. (2017). Exploring the school dropout factors among indigenous students in Melaka. *Journal of Administrative Science Special Edition: Socio-Economic Issue, 14*(3), 1–13.

Sjoberg, S. V. E. I. N., & Schreiner, C. (2003). ROSE: The relevance of science education: Ideas and rationale behind a cross-cultural comparative project. In *4th Conference of the European Science Education Research Association (ESERA):*

Research and the Quality of Science Education. Noordwijkerhout, The Netherlands (August 19–23). En. http://www1. phys. uu. nl/esera2003/program. shtml

Sleeter, C. E. (2012). Confronting the marginalization of culturally responsive pedagogy. *Urban Education, 47,* 562–584.

Snively, G., & Corsiglia, J. (2001). Discovering indigenous science: Implications for science education. *Science Education, 85*(1), 6–34.

Warren, B., Ballenger, C., Ogonowski, M., Rosebery, A. S., & Hudicourt-Barnes, J. (2001). Rethinking diversity in learning science: The logic of everyday sense-making. *Journal of Research in Science Teaching: The Official Journal of the National Association for Research in Science Teaching, 38*(5), 529–552.

Wilson, B. (1981). The cultural contexts of science and mathematics education: Preparation of a bibliographic guide. *Studies in Science Education, 8,* 27–44.

Yilmaz, K. (2008). Social studies teachers' views of learner-centered instruction. *European Journal of Teacher Education, 31*(1), 35–53.

5 The Role of Funds of Knowledge in Culturally Responsive Science Pedagogy

Mohd Norawi Ali and Hartini Hashim

Introduction

Educational stakeholders across the globe are demanding for science education reform that attends simultaneously to the needs of culturally diverse students and promotes academic excellence (Brown & Crippen, 2016). Educators need to take the initiative to build connections with students and their families to make the learning more engaging, motivating, memorable and meaningful. Even though this effort is often not the top priority for most educators, knowing the cultural experience and environmental backgrounds of the students would give advantages that would enliven the learning process.

Every student, regardless of geographical, ethnic and racial boundaries, has his/her own unique cultural experience. Students come to school bringing a lot of their cultural knowledge, skills, and experiences as a result of the cultural practices and interactions in their family and community that occur naturally. Their role and involvement in their family, school and community make them experts in specific areas and skills. These cultural experiences are very valuable and should be tapped by educators by associating and linking these experiences with classroom teaching and learning activities (Mccollough, 2019). Tapping and linking these cultural experiences in classroom activities are also seeing as an act of teachers being Culturally Responsive in their pedagogy (CRSP).

The cultural experiences are referred to funds of knowledge (FoK). Several positive effects of using students' FoK have been described. The integration of FoK in the classroom can improve the relationship between teachers and students because the teachers know the students in a wider context than just the classroom and because there is a more active and personal communication between teachers and students (Barton & Tan, 2009; Irvine, 2003). Apart from that, it will encourage students to actively contribute and participate in the learning process, bridging the gap between their personal experiences and the formal curriculum, therefore enhancing academic learning (Barron et al., 2021; Subero et al., 2017).

According to FoK theory, teachers can bridge the gap between school and home by drawing on the knowledge and skills that students acquire in their families and communities, thereby supporting academic learning (González et al., 2005; Hogg, 2011). The FoK approach was introduced and developed

DOI: 10.4324/9781003168706-6

with the aim of bridging the gap between home and school while avoiding deficit theorising (Gilde & Volman, 2021).

What makes the learning environment so meaningful and lively is that FoK are culturally relevant to students. It gives the chance to students with diverse sets of expertise to engage in the classroom. Indirectly, it provides the opportunity for teachers to internalise and understand students' lives. This allows the teacher to create a bridge between students' home cultures with the science curriculum (Bouillion & Gomez, 2001; Saathoff, 2015).

To fully maximise students' FoK, pedagogy that emphasises cultural responsiveness needs to be creatively planned to bridge the gap of students' cultural diversity. Culturally responsive pedagogy (CRP) that uses cultural modelling as a framework provides equitable and accelerated learning opportunities for all students (Azam & Goodnough, 2018) while empowering students whose culture and language may not be visible in the classroom (Gay, 2010).

Funds of Knowledge from the Anthropology Perspective

Everyone has a cultural experience that exists collectively because of interactions with family members and the surrounding community. The study about FoK involves the studies of culture or cultural experiences and practices that are resources for the survival of a family, community and society (Moll et al., 1992).

When investigating FoK, we need to look at it from the point of view and thinking of an anthropologist, where a person's way of life, how they know something, how they do things and how they build relationships are observed. In doing so, the collected data comes from outside of the classroom, and is generally quite extensive. For example, much of the initial research on FoK involved visits to homes and other community places to understand how and what people know. The information obtained is very useful to be adapted in the curriculum by linking home, community and school. From the anthropology perspective, FoK emphasise communities or family units as holders of historical and cultural knowledge. Therefore, this can dramatically change as the community or family adjusts to the new situation.

Moll and Greenberg (1990) articulate FoK as the clusters of skills and knowledge acquired through cultural and historical interactions and are important for an individual to function appropriately within their community. The concept of FoK is based on a simple premise: people are competent, they have knowledge and their life experiences have given them that knowledge (González et al., 2005). Meanwhile, Beth et al. (2001) and Basu and Calabrese Barton (2007) regard FoK as a cultural and historical knowledge or specific experiences in the family and community contexts. This includes knowledge, action, disposition and behaviour, with the recognition that each domain is culturally shaped. According to these researchers, the concept of FoK will give recognition to the life experiences of an individual in the family or community, which will produce useful, effective and transferable knowledge. Lee and Fradd (1998) further described cultural experience as the students' experience or existing knowledge of various cultural objects and artefacts that are commonly found in their surroundings and communities which are customary to them.

Moll et al. (1992) in their study of multi-culture Hispanic students found that their students have rich and diverse cultural knowledge and cultural experience as a result of mixed and active involvement in a multi-ethnic community. According to Moll et al., teachers' failure to link the resources of the students' cultural knowledge in the teaching and learning process will lead to didactic ways of teaching that make the classroom climate dull and less engaging. Furthermore, the discontinuity between school and home can result in students, particularly those with an ethnic minority or lower socio-economic background, losing their interest in school and performing below their abilities (Bronkhorst & Akkerman, 2016).

In a nutshell, educators need to take the initiative to observe the way of life, practices and culture of students with their community, how they interact together and perform an activity and the changes that occur according to the latest situation so that it can be adapted to the classroom curriculum.

Funds of Knowledge and Science Learning

Although many studies report that students are less interested in science subjects because the subjects' lessons and contents do not relate to their interests and experiences, little research has been conducted on the relationship of personal experience in maintaining the interest and motivation of students in science (Basu & Calabrese Barton, 2007; Othman Talib et al., 2009). Among the researchers who have studied the area include (Atwater, 1996) and (Hammond, 2001) who focused on the role of students' experience, background, interest, skills and knowledge as a strategy in teaching to make science teaching more culturally relevant.

The advantages of exploiting FoK in science teaching and learning to maintain the interest, self-efficacy, motivation and understanding of students of science concepts have been explored by several researchers (see e.g. Azman, 2009; Basu, 2008; Bouillion & Gomez, 2001; Hammond, 2001; Mohd Norawi, 2014; Seiler, 2001). In his garden project study, Hammond (2001) used the community's FoK to generate science-affiliated topics. As a result of his research, a multi-disciplinary discipline has been established that can be accessed by all collaborators based on school standards. Meanwhile, Seiler (2001) in his study assessed the use of students' FoK to discuss science concepts during lunch club activities, and his findings showed that students in the city have extensive scientific knowledge.

Through the concept of community-sharing, opportunities for knowledge exchange bi-directionality between schools and various community groups to help students see science learning in more meaningful contexts were spearheaded by (Moll & Greenberg, 1990) and (Moll et al., 1993) in their study which focused on the relationship between school and home in exchanging FoK. Similarly, Upadhyay (2005) used a case study method to investigate a female teacher who taught basic science to students in urban schools that have social, cultural, linguistic and ethnic diversity.

Ali et al. (2017) used the experience of students visiting a night market held over the weekend to explain the concept of different medium density in science teaching for the topic of light and refraction. Meanwhile, Azman (2009) in his study among low achieving students in a rural school used the cultural experience of the students while picnicking at a waterfall near the students' village to create a conducive learning environment to explain the concepts of velocity, acceleration and distance. Rohandi (2010) in his study involving secondary students in Indonesia activated the students' learning from artefacts in the community such as guitars and cultural songs to explain the formation of rainbows. These situations illustrate the enthusiasm and creativity of educators in utilising contextual knowledge that are abundant in the students' environments and community to ease understanding of science concepts.

Ahmad Nurulazam Md Zain et al. (2010) in their research on the integration of FoK in instructional congruence during classroom science teaching in three low-performing schools in Penang, Malaysia, found a positive impact on students' practical aspects of science work, out-of-school science, future participation in science and a combination of overall attitudes and interests in science subjects. This indicates that learners with lower cognition gain benefit through the blending of FoK with their science subjects or lessons if teachers are creative and explicit in teaching.

The importance of FoK therefore becomes clear as it ensures that all students can pursue science subjects on an equitable basis regardless of geographical boundaries. FoK is a strategic and cultural resource that could be a useful asset in the classroom to help teachers understand the cultural system of minority students (Brown & Crippen, 2016; Moll et al., 1992). Nonetheless, not all of the students' cultural experiences can be integrated during science teaching in the classroom.

Funds of Knowledge of Diverse Students

Teachers' consideration of the cultural experiences of the students' life can create a harmonious and active learning environment. Indirectly, it will provide an opportunity for the correlation between the school science curriculum and students' cultural experiences with their communities. Upadhyay (2005) suggested that teachers should take the initiative to deepen the contextual knowledge of the students which is contained in their knowledge treasures to be integrated into the teaching process to produce meaningful learning.

Teachers can help students in the learning process by using what students already know. This is because knowledge schemas will evolve when students are able to relate new information or concepts to things they already know or experience. Students learn a new science concept better when they are given the opportunity to share their expertise as well as existing knowledge related to the concept introduced by the teacher.

Apart from the need for teachers to have awareness and knowledge of the FoK of students, they also need to have an open mind in implementing the responsive cultural pedagogy to meet the diverse needs of students. Teachers' readiness means they need to be active and diligent in associating the element of FoK with the science curriculum creatively through learning activities that are relevant to students' culture and daily experiences.

Strategy to Tap Students' FoK

Teachers need to take the initiative to visit students' home, garden or village to see for themselves the living culture and objects found in the students' environment even though this is not their primary task. Teachers can discover students' FoK in a variety of ways, as evidenced by the exemplary practices, action plans, logbooks and interviews (Judith & Volmen, 2021). Students' FoK could be discovered through the context of a specific course or theme, and classroom discussions are particularly useful in this regard.

To tap students' FoK, teachers basically need to interact with the students and their families by engaging in conversations about their home life and interests. Sometimes, teachers may need to do home visits to gain better and greater understanding of the many aspects of the students' life and the types of knowledge they hold which they could integrate into their classroom. Following their home visits, teachers can design new curriculum and lesson plans based on the FoK they have identified (Denton & Borrego, 2021).

Parent involvement has been a subject of academic interest for decades, and there is increasing recognition that race, ethnicity, socio-economic group and gender are privileging or deprivileging conditions that affect how parents are positioned in school (Allen & White-Smith, 2018; Blackmore & Hutchison, 2010; Shuffleton, 2017). Teachers can also invite parents or guardians to the classroom to share their experiences and expertise with the students. Some parents may not believe they have something to share but knowing that they do will make the sharing easier. For example, a teacher can ask a student's mother to share her experience in the market if the teacher knows that the student's mother is a seller of vegetables in the market. In addition, teachers can invite parents to share their knowledge, experiences and skills outside the classroom timetable, during club activities or on weekends.

Teachers also need to be open-minded by giving students opportunities to share their cultural experiences. By doing so, students in minority groups who are often marginalised will have the opportunity to showcase themselves by sharing their special skills and abilities with their peers and teachers. This sharing of opportunities will build students' scientific communication skills and self-confidence.

The knowledge gained from students' families, whether from home visits or parents coming to school to share their expertise as a way to connect with their children, can be very useful. By incorporating a little bit of important knowledge of each student's family in the classroom, it would certainly allow teachers to interact with each student more warmly. Students will feel valued when their ideas and expertise are considered.

The following description provides an approach that suits the environment of secondary students. Teachers can use the experience of students involved in the preparation of rice and curry dishes for guests during a wedding feast. Many science concepts in addition to science process skills can be learned. The teacher can start, for example, by asking the students to measure rice and water following a certain ratio. For instance, students can be asked to measure the rice with water according to a certain rate of 1 part rice and 1½-part water to get a perfectly cooked rice. Ask the students to position the height of the pot appropriately on the campfire so that it cooks quickly. The teacher can explain that when cooking rice or curry in a large pot, the campfire is lit with a strong flame at first, and the rice is stirred so that it cooks quickly. When it is almost cooked, the fire is reduced, and the pot is covered with a lid so that the rice is perfectly cooked.

Teachers can take advantage of these opportunities to allow students to share their cultural experiences and expertise in front of the class to increase their confidence and self-efficacy. For example, after a visit to the zoo, the teacher can discuss with the students about the ways animals feed and their habit of hiding food. Students can act as a zoologist in a scenario and explain to other students about animals and crops eaten by animals which can either be carried out spontaneously or planned. In other scenarios, teachers can ask students to play a role appropriate to their FoK in activities that can either be carried out spontaneously or planned in advance.

Additionally, students' FoK can range from small insects, trees, fertilisers, religious activities, carpentry, computers, iPads, speaking foreign languages, writing poetry and songs to designing vehicles and atomic bomb and development. Teachers therefore need to have openness and a little sacrifice of time to evaluate and examine the strengths and uniqueness of each student that can be applied in the learning activities.

Types of Funds of Knowledge

FoK can include knowledge of how to grow vegetables in a limited urban space. Gardening is considered one kind of FoK where children most often acquire this knowledge from their parents and other family members (Cun, 2020). FoK can also include ways to make special drinks like *teh tarik* or famous Malaysian food like *roti canai* and *sate*. Additionally, it could include knowledge like methods of arranging cars in confined spaces, how to control traffic in crowded areas such as night markets, how to determine the quality of fruits such as *durians* or *rambutan*, or how to raise animals such as chickens, goats or cows. The same goes for the halal and acceptable way of slaughtering animals based on religious views. FoK could also be a way of designing and installing banners to attract attention, and various kinds of other knowledge that students gain from their interactions with family and the community. Moll (2014) in his study found that some existing knowledge from students' home practices such as knowledge of agriculture, household management and religion are knowledge that schools and teachers did not know students possessed.

A lot of knowledge and skills can be gained through interaction with the family at home such as while cooking rice, frying fish, making drinks or juices, making salads, washing dishes or growing vegetables hydroponically during the hardship of the COVID-19 pandemic. The same happens in the community that involves various activities such as playing football, jogging, hill climbing, swimming in the sea, volunteering to clean certain areas or places, and helping those in need.

Components of Funds of Knowledge

FoK comprises three main elements: source, process and content as summarised in Table 5.1.

Hogg (2011) identified that sources contributing to students' FoK include but are not limited to families, peers, popular culture, community and lived experiences. Social interaction is the key in developing FoK; therefore, exploring these interactions could be the means to identify FoK sources.

According to Vélez-Ibáñez and Greenberg (1992), research on the formation and transformation of FoK among Mexican American households highlights that the FoK transmission conditions may support students' learning in schools. Adult family members organise learning within a "zone of comfort" so children can learn within an encouraging environment where inquiring as they practice is a good skill while making mistakes is a part of learning. Yet, the knowledge and skills that students gain in this transmission process may not be apparent. Instructors can help students think through the assets that they have gained through these processes (Rios-Aguilar et al., 2011) so that learning becomes more engaging and meaningful.

FoK content encompasses the accumulated treasures of knowledge and skills that students acquire as a result of their participation in social interactions. It is important to note that FoK research conceptualises culture as fluid, undergoing change over time and existing through social interaction. For example, the FoK found in Mexican American households include knowledge of labour law; folk medicine; design and architecture (Moll et al., 1992); transnational knowledge; bilingual and cultural skills and knowledge; agriculture; and international trade (González et al., 2005). Additionally, Lew (2009) stressed that although knowledge funds are triggered by challenging life experiences such as discrimination and violence, these can provide valuable knowledge and skills. Recently, Neri (2018) articulated that students' FoK in college police programmes include knowledge of how to interact with the police, how police treat homeless people, police priorities and group violence.

Table 5.1 Components of funds of knowledge

Source	Process	Content
The person, place or thing that is part of the social interaction and from where the FoK is obtained.	The actions or steps that facilitate the transmission of knowledge and skills.	The area and substance of knowledge and/or skills.

Cooking, Food and Beverage

It is evident that some students gain cooking knowledge by observing their mother and grandmother. Thus, even though they have never tried cooking the dish independently, they would be able to display their knowledge by describing the ingredients whenever asked. The experience of making hot drinks such as *teh tarik*, coffee or Nescafe or cold drinks from various types of fruits or *cendol* can be associated with the topic of compounds, solutions, solutes and solvents. When hot drinks are mixed with ice, the temperature of the final solution will be reduced because of the occurrence of heat loss by hot water due to absorption by ice and the environment. Hence, the concept of thermal equilibrium that exists can be easily explained using students' FoK.

A nutritious and healthy culinary product depends on a balanced diet consisting of various food classes and quality, in addition to the price and size of ingredients for cooking as well as cooking techniques such as frying, boiling or steaming or grilling. Additionally, the use of the type of cooking utensils and the way the energy sources whether gas or electricity is used also determines the monthly costs that need to be borne.

Small Home Industry

The fermentation is the basic chemistry process that will produce *tapai*. Ragi is used in making *tapai*. Ragi is a dry-starter culture prepared from a mixture of rice flour, spices and water or sugar from cane juice/extract. It is usually used to ferment cassava and glutinous rice into "*tapai* or tape", a popular Malaysian delicacy, normally consumed as dessert. Other types of food that are produced based on the fermentation process are *kimchi* and *tempeh*.

Kimchi, a staple in Korean cuisine, is a traditional side dish of salted and fermented vegetables, such as napa cabbage and Korean radish, made with a wide selection of seasonings including *gochugaru*, spring onion, garlic, ginger, jeotgal, and many others. It is also used in a variety of soups. Meanwhile, *tempeh* is a traditional Javanese food made from fermented soybeans. It is made using a natural culturing and controlled fermentation process that binds soybeans into a cake form.

Gardening

Gardening activities that involve various types of fruits and vegetables train students to be healthier as well as provide economic resources for the family. Knowledge of growth needs such as water, sunlight, soil medium and fertilisers inherited from parents or learned from the community can build students' confidence to engage in modern agriculture. They can develop skills in hydroponic or aquaponic cultivation that involves the integration of plants and fish.

Sport and Recreational Activities

Many sports and leisure activities provide knowledge, skills and values to students. Cycling activities, for example, allow them to understand the gear system and the concepts of velocity and acceleration. The recreational experience on the beach provides awareness of sea and land breezes and the circulation of day and night.

Transportation and Vehicles

Students' experience repairing toys, bicycles, motorcycles or cars gives them the opportunity to understand the design aspects. They can apply several mechanical concepts such as balance of force, torque, momentum and friction. This wealth of knowledge can motivate students to explore the field of motoring and mechanical engineering. For this reason, students' relationships and interactions with family members, siblings or peers can become a factor for students to venture into a specific field or career path (Hedges & Jones, 2012). Students who grew up in families engaged in mechanical and vehicle repair would often have the creativity to modify motorcycles or cars and would thus be more interested in Physics or engineering.

Language and Song

Parents' interactions with children can help them preserve their mother tongue and cultural heritage (Davila et al., 2017; Durand & Perez, 2013). Traditional songs in the Malay culture of *Dikir Barat* heritage such as *anak tupai* can explain the uniqueness of squirrels. The song *Bangau oh Bangau* can be associated with the concept of ecosystems and biodiversity. Some of the idioms given next can be used as linguistic scaffolding in helping students understand the new science concepts and scientific knowledge and skills introduced to them in the classroom:

1   *Naik minyak* (blow your top)
2   *Biar lambat asalkan selamat* (slow and steady wins the race)
3   *Banyak udang banyak garam, banyak orang banyak ragam* (It takes all sorts to make a world)
4   *Ikut resmi padi, semakin berisi semakin tunduk, Jangan ikut resmi lalang semakin tinggi semakin menegak* (Be like the paddy stalk, it bends low as it is laden with ripening grains, don't be like the tall grass, it stands tall but without seeds)

Festivals and Celebrations

The experience of students celebrating special days such as *Hari Raya Aidil Fitri*, *Hari Raya Korban*, Chinese New Year, Deepavali and *Gawai* in the context of Malaysia provides opportunities for students (and teachers) to get to know the culture of the plural society that lives in harmony and mutual tolerance in Malaysia.

In addition, the experience of fasting in the month of Ramadan can provide awareness of the importance of health aspects and a balanced diet. Similarly, students who are involved in sacrificial slaughter will better understand the digestive, respiratory and circulatory systems in animals, in addition to understanding the position and function of the internal organs of animals.

During the Chinese New Year celebration, the Chinese community does not miss making the traditional glutinous rice cake known as *Kuih Bakul*. The process of making the cakes with a mixture of certain ingredients and steaming techniques can be linked to the learning of science related to the topic of properties of materials.

Phenomenological Event

Natural phenomena such as lunar eclipses, solar eclipses, changes in the shape of the moon, sea tides and rivers can be used in the teaching of science while giving awareness of the greatness of God Almighty, the Creator of the universe. Over the past few decades, the earth has experienced rapid warming (New et al., 2009). Human activities are the primary drivers of climate change as they contribute to more than 95% of the rapid temperature rise (Cook et al., 2016) especially activities like the burning of fossil fuels, deforestation and land-use changes that emit greenhouse gases. In response to this, indigenous people and their traditional ecological knowledge have gained more attention because of their ability to address climate change at the grassroots level (Kupika et al., 2019). We need local knowledge of the environment to detect changes in the local surroundings such as the flood season which is affecting the Western region of Malaysia more and more recently.

Cultural Objects and Artefacts

There are a variety of cultural objects and artefacts commonly found in the students' environment that can be activated in learning. Cultural objects consist of shops, markets, houses, houses of worship, buildings, roads, rivers, seas, schools, clinics, hospitals, airports, farms, plantation areas, industries, weapons and, musical instruments. Apart from that, cultural objects include common areas or locations visited, tourism centres, picnic or recreational spots, shopping complex, vehicles and so on.

Religious Practices

Students' religion and their practices would be an interesting topic to discover in teachers' efforts to build rapport with them. Like other cultural practices, religious knowledge and practices are passed down from generation to generation. Praying and reading religious texts such as the Quran are literacy practices that many families engage in daily. Additionally, the findings from the Cun (2020) study revealed that religious practices and literacy practices support the maintenance of the home language and the family's harmonious relationship.

Incorporating FoK in Culturally Responsive Pedagogy

In providing equitable education across geographical boundaries, understanding the fund of knowledge framework is vital in making science teaching culturally relevant to each student. After knowing the ways of how FoK and its types can be tapped, educators need creativity to apply it intelligently in CRP to elicit an effective resonant effect on students' engagement and understanding.

Gay (2010) described CRP as the use of diverse students' cultural knowledge, prior experiences, frames of reference and performance styles to make learning encounters more relevant and effective for the students. Gay's experience shows that when academic knowledge and skills are taught within the lived experiences and frames of reference of students, they are more personally meaningful, have higher interest appeal and are learned more easily and thoroughly. In CRP, the teaching and learning process focuses on the cultural experiences and backgrounds of students. According to Garcia (1996), CRP comprises six essential characteristics known as 5Rs and 1T, which include (i) respect, (ii) responsiveness, (iii) responsibility, (iv) resourcefulness, (v) reasonableness and (vi) theory, in order to make teaching inclusive for all students.

R1 refers to *respect* which means showing respect for the culture and way of life of the students as well as considering the experience, background knowledge and worldview brought by the students into the class. Respect also refers to the teacher's recognition of certain experiences, ideas and abilities that students have because of interactions with family and community. This situation will create a sense of belonging that can encourage and motivate them to come to school and be actively involved in the lessons and learning activities. Through respect, a more harmonious relationship between teachers, students and the community will be built as students' expertise in certain fields is acknowledged and appreciation of the students' culture is promoted.

R2 refers to *responsiveness* which is the ability of teachers to understand and familiarise themselves with the culture, FoK and worldview of their students and subsequently make appropriate adaptations in teaching strategies to ensure that students have a meaningful and engaging learning experience. The ability of teachers to be responsive towards students' culture such as their mother tongue would be key to engaging students from diverse cultural and linguistic backgrounds (Azam & Goodnough, 2018). In the implementation, teachers will adapt the instruction, curriculum or teaching content to the cultural backgrounds of the students by taking their differences in prior knowledge, experiences, readiness, language culture and interests into account (Alhanachi et al., 2021).

R3 refers to *responsibility* which demands a teacher's awareness of the short-comings and the achievement gaps that their students may be experiencing while understanding the causes that contribute to these gaps and demonstrating effective teaching to minimise them. From the CRT's perspective, responsibility also demands staying away from "deficit models" and avoiding lowering of standards for diverse student populations.

R4 refers to *resourcefulness* which includes gaining access to financial, logistical and intellectual resources that teachers can use to engage students in learning that may ultimately help students to succeed. García (2005) evaluated intellectual resources more highly than others and considered financial resources as simply providing professional development for teachers so that they may have more access to intellectual resources to teach their culturally diverse students. Additionally, resourcefulness also refers to the ability and creativity of a person to face difficulties by using all the wisdom and what is available or around at that time to overcome the problems faced.

R5 refers to *reasonableness* which requires a teacher to be realistic and reasonable with their expectations, teaching strategies and assessment tools. A reasonable teacher will plan the curriculum experience carefully and thoughtfully so that it can be understood and responded to by a diverse range of students.

T1 refers to theory which includes "conceptual and theoretical approaches" to learning and teaching to engage all students. Research-based pedagogies should be considered for planning and teaching, resulting in science teachers becoming facilitators of learning while using student-centred, inquiry-oriented approaches to teaching. Learning is most effective when new knowledge and skills are linked to existing knowledge socially and the individual builds meaning for oneself.

Exemplar Association of FoK in Culturally Responsive Science Pedagogy

After critically examining the resources that are available in the homes or communities, the original intent of FoK work involves teachers planning themed units that emerged from the home analyses. These themes, then, require students to draw on knowledge from outside of the classroom, with people and things they are familiar with. The students need to actively participate and engage in inquiry as well. This is because the themes are something they have access to. The teacher would need to be able to actively fit the "school stuff" into the themes.

For example, the unit on gardening theme can be appropriately applied to describe elements of science, mathematics, technology and environmental knowledge. Perhaps, students' experiences in the environment outside the classroom, together with parents who work in gardens, or even experiences of visits to agricultural parks and farmers markets nearby can be sources of FoK that can be exploited. During the COVID-19 pandemic, by selecting one type of FoK regarding home gardening, teachers can use a STEM project-based approach that adapts the 5R-1T elements in the lesson plan (see Table 5.2).

Outline of the Lesson Plan

STEM Project:　　　　　　Building a model for urban farming in the backyard
Class/School Level:　　　Secondary School
Topic:　　　　　　　　　Plant Growth
Skills:　　　　　　　　　(a) Critical/Creative thinking
　　　　　　　　　　　　(b) Problem solving
　　　　　　　　　　　　(c) Decision making
　　　　　　　　　　　　(d) Interpersonal communication
Prerequisite Knowledge:　Requirements for plant growth
Students' FoK include:

- Students are always helping the family to plant vegetables, flowers and fruits.
- Students often make compost fertilisers from household waste and faeces of livestock such as chicken, goats and cows.
- Students have experience buying and selling vegetables at farmers' markets and night markets.

Table 5.2 Adapting funds of knowledge in culturally responsive science teaching

CRP Component (5R + 1T)	Teacher's Concern	Learning Activities	Script/dialogue used by the teacher to connect students' funds of knowledge	Skills
Respect	Acknowledging students' expertise in certain fields.	Teacher explores students' contextual knowledge by asking questions.	• Do you like green surroundings? • Have you ever grown vegetables or flowers in your backyard? • Do you have experience buying vegetables from the market? Or do you go with your parents to buy vegetables at the market or shop?	Critical thinking
Responsiveness	Understanding and familiarising with the culture, FoK and worldview of the students.	Teacher asks students who have experience in planting flowers or vegetables. Students are required to describe how they plant the flowers or vegetables.	• I realise that most of you have experience in growing flowers or vegetables. Can you explain where you plant it? In the flowerpot or directly on the ground?	Critical thinking
Responsibility	Awareness of the weaknesses and achievement gaps that their students may be experiencing while understanding the causes that contribute to these gaps and demonstrating effective teaching.	Students explain how they or their parents plant the flowers if they live in flats or apartments.	• Maybe some of you live in different areas and plant different crops? • What kinds of used materials or household waste can be used for making fertilisers?	Problem solving

CRP Component (5R + 1T)	Teacher's Concern	Learning Activities	Script/dialogue used by the teacher to connect students' funds of knowledge	Skills
Resourcefulness	Gaining access to financial, logistical and intellectual resources that the teacher can use to engage students and eventually help them to be successful.	Students create a mini garden model in the backyard using materials available.	• Why don't we try to plant vegetables today? You may use any recycled bottles as plant pots or a self-watering system. • How do you customise the urban farming model by using existing waste materials such as bottles, plastic, PVC pipes and others? • For those students who live in a flat or an apartment, they can make it simple and place it on the balcony or in their kitchen.	Creative thinking
Reasonable ness	Being realistic and reasonable with expectations, teaching strategies and assessment tools.	Students share the photos of the progress of their project.	During the pandemic, we are not encouraged to go out. So, planting your own vegetables helps you from having to go out and can also help your family save money. It also provides green and healthy products for your family's consumption.	Problem solving and decision making
Theory	Applying the social constructivist theory and inquiry models.	Students are encouraged to use their experience and skills. Collaborate with friends to build an urban farming model.	Can you share the building design of your model garden?	Interpersonal communication

Conclusion

As a conclusion, it can be said that FoK as an element of contextual knowledge must be taken into consideration by science educators in classroom teaching and even during co-curricular activities, in order for the science curriculum to have relevance with the cultural experience of students in their families, communities and environments. Students will be thrilled when their cultural experience is recognised and highlighted by their teachers in the classroom. Science education needs to look at students' cultural knowledge, background and ways of knowing to produce a pedagogy that is inclusive and equitable for diverse students. Through creatively integrating FoK in the culturally responsive science pedagogy, it would allow teachers to have a more holistic view of students and their families that leads to intimate relationships that benefit students and the teaching practice. When science curricula, academic knowledge and skills are taught within students' life experiences as well as cultural and reference frameworks, they are personally more meaningful, have a higher interest appeal, and are more easily and comprehensively learned. Additionally, the concepts learned in the subject of science need to be reinforced in other subjects such as language, mathematics, technology, religion, designs and artistry, so that the students can see that the concepts are associated with the reality of their lives.

References

Ahmad Nurulazam Md Zain, Rohandi, & Jusoh, Azman. (2010). Instructional congruence to improve Malaysian students' attitudes and interest toward science in low performing secondary schools. *European Journal of Social Sciences, 13*(1), 89–100.

Alhanachi, S., de Meijer, L. A. L., & Severiens, S. E. (2021). Improving culturally responsive teaching through professional learning communities: A qualitative study in Dutch pre-vocational schools. *International Journal of Educational Research, 105*, 101698. https://doi.org/10.1016/j.ijer.2020.101698

Ali, M. N., Halim, L., Osman, K., & Mohtar, L. E. (2017). The integration of fund of knowledge in the hybridization cognitive strategy to enhance secondary students' understanding of physics optical concepts and remediating their misconceptions. In M. Karpudewan, A. N. Md Zain, & A. L. Chandrasegaran (Eds.), *Overcoming students' misconceptions in science: Strategies and perspectives from Malaysia* (pp. 181–201). Springer.

Allen, Q., & White-Smith, K. (2018). That's why I say stay in school: Black mothers' parental involvement, cultural wealth, and exclusion in their son's schooling. *Urban Education, 53*(3), 409–435. https://doi.org/10.1177/0042085917714516

Atwater, M. M. (1996). Social constructivism: Infusion into the multicultural science education research agenda. *Journal of Research in Science Teaching, 33*(8), 821–837.

Azam, S., & Goodnough, K. (2018). Learning together about culturally relevant science teacher education: Indigenizing a science methods course. *International Journal of Innovation in Science and Mathematics Education, 26*(2), 74–88.

Azman, J. (2009). *Kesan strategi padanan instruksi ke atas persekitaran pembelajaran, konsep kendiri akademik dan efikasi kendiri pelajar Fizik* [*Effect of instructional congruent on learning environment, academic self-concept and self-efficacy of physics students*] (PhD Thesis). Universiti Sains Malaysia, Penang.

Barron, H. A., Brown, J. C., & Cotner, S. (2021). The culturally responsive science teaching practices of undergraduate biology teaching assistants. *Journal of Research in Science Teaching, 58*(9), 1320–1358. https://doi.org/10.1002/tea.21711

Barton, A. C., & Tan, E. (2009). Funds of knowledge and discourses and hybrid space. *Journal of Research in Science Teaching, 46*(1), 50–73.

Basu, S. J. (2008). How students design and enact physics lessons: Five immigrant Caribbean youth and the cultivation of student voice. *Journal of Research in Science Teaching, 45*(8), 881–899.

Basu, S. J., & Calabrese Barton, A. (2007). Developing a sustained interest in science among urban minority youth. *Journal of Research in Science Teaching, 44*(3), 466–489.

Beth, W., Cynthia, B., Mark, O., Ann, S. R., & Josiane, H.-B. (2001). Rethinking diversity in learning science: The logic of everyday sense-making. *Journal of Research in Science Teaching, 38*(5), 529–552.

Blackmore, J., & Hutchison, K. (2010). Ambivalent relations: The "tricky footwork" of parental involvement in school communities. *International Journal of Inclusive Education, 14*(5), 499–515.

Bouillion, L. M., & Gomez, L. M. (2001). Connecting school and community with science learning: Real world problems and school-community partnerships as contextual scaffolds. *Journal of Research in Science Teaching, 38*(8), 878–898.

Bronkhorst, L., & Akkerman, S. (2016). At the boundary of school: Continuity and discontinuity in learning across contexts. *Educational Research Review, 19*.

Brown, J. C., & Crippen, K. J. (2016). Designing for culturally responsive science education through professional development. *International Journal of Science Education, 38*(3), 470–492. https://doi.org/10.1080/09500693.2015.1136756

Cook, J., Oreskes, N., Doran, P. T., Anderegg, W. R. L., Verheggen, B., Maibach, E. W., Carlton, J. S., Lewandowsky, S., Skuce, A. G., & Green, S. A. (2016). Consensus on consensus: A synthesis of consensus estimates on human-caused global warning. *Environmental Research Letters, 11*.

Cun, A. (2020). Funds of knowledge: Early learning in Burmese families. *Early Childhood Education Journal, 49*, 711–723.

Davila, D., Nougueron, S., & Vasquez-Dominguez, M. (2017). The Latinx family: Learning literature at the library. *The Bilingual Review, 33*(5), 33–49.

Denton, M., & Borrego, M. (2021). Funds of knowledge in STEM education: A scoping review. *Studies in Engineering Education, 1*(2), 71–92.

Durand, T. M., & Perez, N. A. (2013). Continuity and variability in the parental involvement and advocacy beliefs of Latino families of young children: Finding the potential for a collective voice. *School Community Journal, 23*(1), 49–79.

Garcia, E. E. (1996). Preparing instructional professionals for linguistically and culturally diverse students. In J. Sikula (Ed.), *Handbook of research on teacher education* (pp. 802–813). Simon & Schuster Macmillan.

García, E. E. (2005). *Teaching and learning in two languages: Bilingualism and schooling in the United States.* Teachers College Press.

Gay, G. (2010). *Culturally responsive teaching: Theory, research, and practice* (2nd ed.). Teachers College Press.

Gilde, J., & Volman, M. (2021). Finding and using students' funds of knowledge and identity in superdiverse primary schools: A collaborative action research project. *Cambridge Journal of Education, 51*(6), 673–692.

González, N., Moll, L. C., & Amanti, C. (2005). *Funds of knowledge: Theorizing practices in households, communities, and classrooms.* Routledge.

Hammond, L. (2001). Notes from California: An anthropological approach to urban science education for language minority families. *Journal of Research in Science Teaching, 38*, 983–999.

Hedges, H., & Jones, S. (2012). Children's working theories: The neglected sibling of the learning outcomes. *Early Childhood Folio, 16*.

Hogg, L. (2011). Funds of knowledge: An investigation of coherence within the literature. *Teaching and Teacher Education, 27*(3), 666–677. https://doi.org/10.1016/j.tate.2010.11.005

Irvine, J. J. (2003). *Educating teachers for diversity: Seeing with a cultural eye.* Teacher College Press.

Kupika, O. L., Gandiwa, E., Nhamo, G., & Kativu, S. (2019). Local ecological knowledge on climate change and ecosystem-based adaptation strategies promote resilience in the Middle Zambezi biosphere reserve, Zimbabwe. *Scientific, 2019*, 1–15. https://doi.org/10.1155/2019/3069254

Lee, O., & Fradd, S. H. (1998). Science for all, including students from non-English language backgrounds. *Educational Researcher, 27*(4), 12–21.

Lew, Z. (2009). Dark funds of knowledge, Deep funds of pedagogy: Exploring boundaries between lifeworlds and schools. *Discourse: Studies in the Cultural Politics of Education, 30*(3), 317–331.

McCollough, C. (2019). Reforming science teacher education with cultural reflection and practice. *International Journal of Learning, Teaching and Educational Research, 18*(1), 31–49. https://doi.org/10.26803/ijlter.18.1.3

Mohd Norawi, A. (2014). *Pembangunan dan Keberkesanan Modul Kereta Solar dalam Projek Berasaskan Sains terhadap Motivasi dan Pemikiran Inovatif Pelajar Sains* [*Development and effectiveness of solar car innovation module in project based science toward nurturing motivation and innovative thinking of science students*] (PhD Thesis). Universiti Kebangsaan Malaysia, Bangi, Selangor.

Moll, L. C., Amanti, C., Neff, B. D., & Gonzalez, N. (1992). Funds of knowledge for teaching: Using a qualitative approach to connect homes and classrooms. *Theory into Practice, 31*(2), 132–141.

Moll, L. C., Tapia, J., & Whitmore, K. F. (1993). Living knowledge: The social distribution of cultural resources for thinking. *Distributed Cognition, 139–163*.

Moll, L. S. (2014). *Vygotsky and education.* Routledge.

Moll, L., & Greenberg, J. (1990). Creating zones of possibilities: Combining social contexts for instruction. In L. Moll (Ed.), *Vygotsky and education: Instructional implications and applications of socio-historical psychology* (pp. 319–348). Cambridge University Press.

Neri, R. C. (2018). Learning from Students' career ideologies and aspirations: Utilizing a funds of knowledge approach to reimagine career and technical education. In J. M. Kiyama & C. Rios-Aguilar (Eds.), *Funds of knowledge in higher education: Honoring students' cultural experiences and resources as strengths.* Routledge.

New, M., Liverman, D., & Anderson, K. (2009). Mind the gap-policymakers must aim to avoid a 2C temperature rise, but plan to adapt to 4C. *Nature Reviews Clinical, 3*.

Othman Talib, Luan W. S., Azhar, Shah Christirani, & Abdullah, Nabilah. (2009). Uncovering Malaysian students' motivation to learning science. *European Journal of Social Sciences, 8*(2), 266–276.

Rios-Aguilar, C., Kiyama., J., Gravitt, M., & Moll, L. (2011). Funds of Knowledge for the poor and forms of capital for the rich? A Capital approach to examining funds of knowledge. *Theory and Research in Education, 9*(2), 163–184.

Rohandi. (2010). *Incorporating student's funds of knowledge to develop a sustained interest in science* (PhD Thesis). Universiti Sains Malaysia, Penang.

Saathoff, S. D. (2015). Funds of knowledge and community cultural wealth: Exploring how pre-service teachers can work effectively with Mexican and Mexican American students. *Critical Questions in Education, 6*(1), 30–40.

Seiler, G. (2001). Reversing the "standard" direction: Science emerging from the lives of African-American students. *Journal of Research in Science Teaching, 38,* 1000–1014.

Shuffleton, A. (2017). Parental involvement and public schools: Disappearing mothers in labor and politics. *Studies in the Philosophy of Education, 36*(21), 21–32.

Subero, D., Vujasinović, E., & Esteban-Guitart, M. (2017). Mobilising funds of identity in and out of school. *Cambridge Journal of Education, 47*(2), 247–263.

Upadhyay, B. R. (2005). Using students' lived experiences in an urban science classroom: An elementary school teacher's thinking. *Science Education, 90*(1), 94–110.

Vélez-Ibáñez, C. G., & Greenberg, J. B. (1992). Formation and transformation of funds of knowledge among U.S.-Mexican households. *Anthropology & Education Quarterly, 23*(4), 313–335. www.jstor.org/stable/3195869

Part II
Case Studies

6 Case Studies in Indonesia

From Cultural Pluralism to Culturally Responsive Science Pedagogy

Murni Ramli

Introduction: What Is Cultural Pluralism in the Context of Indonesia?

Cultural pluralism is an idea or concept which has been discussed since long ago as a term used to figure out unique groups – mostly small groups – within the larger society, in terms of how they apply and keep their uniqueness in terms of values, indigenous knowledge, norms, and lifestyle, and others in which these values and practices are accepted by the dominant majority. Various definitions of cultural pluralism have been introduced by scholars including "a social condition in which communities of different cultures live together and function in an open system" (Pantoja et al., 1976).

Richard J. Bernstein in his writing, "Cultural Pluralism" in Philosophy and Social Criticism, stated that the term *cultural pluralism* was popularised by Horace Kalen who was a student of William James. Together with John Dewey, William James is a proponent of cultural pluralism in the United States, in the era of the early twentieth century. More than 27 million immigrants came to America between 1870 and 1920, and most of them came from southern and eastern Europe (Bernstein, 2015). The context of cultural pluralism in the United States can be said to be the culture brought by immigrants, and how that culture developed in the United States, which in the twentieth century was generally a White Anglo-Saxon Protestant.

Culture in this context has a broad meaning, relating to habits, rituals, perceptions and perspectives, ways of life, as well as knowledge and technology that are used by certain communities daily in supporting their lives. In a country or region, cultural differences are commonplace in society. Different cultures can be attributed to geographical aspects which affect the lifestyle of a tribe or group and are formed from the process of adaptation to the natural conditions in which they live. Additionally, it can also be caused by the interaction between indigenous people and immigrants, as is the case of immigrants in the United States.

Bernstein also argued that a democratic society that respects differences is enriched by these differences which are very relevant in discussing contemporary cultural pluralism in the global context (Bernstein, 2015). Based on this opinion,

DOI: 10.4324/9781003168706-8

cultural pluralism can be seen in terms of how democratic concepts are applied in the world of education, in schools, and in classrooms, which consist of multi-character and multi-ability students. The multi-character and multi abilities possessed by students may be the result of assimilation from their cultural, economic, natural, or geographical backgrounds, and their habits or experiences.

Katherine P. McFarland used two lenses as perspectives of cultural pluralism, that is, multicultural education and critical pedagogy. Multicultural education is a platform in which teachers must work with students from diverse cultural, economic, and language backgrounds. Meanwhile, critical pedagogy is a pedagogical approach where teachers should apply culturally pluralistic classroom activities in which students from various cultural backgrounds get to experience contextual education. Critical pedagogy stresses on the anticipation of social inequalities in the classroom (McFarland, 1999).

Compared to the issue of cultural diversity, which is growing in some immigrant countries, the case in Indonesia is comparatively different. In contrast to the United States, Indonesia is not an area targeted by immigrants from certain areas of the world. However, the indigenous people of the country who inhabit the islands of Indonesia, which was previously called *Nusantara*, are people of different ethnicities. There are approximately 400 indigenous ethnicities in Indonesia with approximately 300 local languages that are still being used by the local communities. Apart from the indigenous tribes on these islands, there are also a small number of immigrants from mainland China, the Arab countries, and India who came to Indonesia, either because of economic factors or for the purpose of religious missions.

Religions are also considered as cultural pluralism in Indonesia (Azra, 2010). There are five religions recognised in Indonesia, namely Islam, Protestant and Catholic Christianity, Buddhism, Hinduism, and Confucianism. A small group of Indonesians also adhere to certain beliefs. Even within one religion, there are various sects and groups, either those born in countries outside Indonesia or ones that emerged in Indonesia. For example, Islam in Indonesia has two large organisations, namely *Muhammadiyah* and *Nahdhatul Ulama*. These two Islamic organisations have different understandings and principles or interpretations on several matters in relation to how Islam is believed and implemented in everyday life. With the existence of religious sects and organisations, the implementation of worship or tradition-based religion in the society is thus becoming diverse.

Based on the two concepts of cultural pluralism proposed by McFarland, that is, multicultural education and critical pedagogy, its application in Indonesia may be very complex and a national standard approach is difficult to be implemented.

Has Cultural Pluralism Been Accommodated in the Science Curriculum and Education System in Indonesia?

The Indonesian education system was adapted from the Western and Eastern systems, and this occurred because of the long period of colonialism experienced by the country. Dutch colonialism which lasted for approximately 350 years led to the thickening of the Dutch or European education system in Indonesia. During the Dutch colonial period, education was divided based on caste or groups in the society, such that the types and quality of schools were divided: at the top were the schools for Europeans or the Dutch, followed by the schools for Indonesian aristocrats, Chinese schools, and Arabic schools, and at the very bottom were schools for the natives and commoners. These types of schools differed not only in terms of their names, but also in terms of the language of instruction used in the schools, the curriculum applied, and the quality and breadth of knowledge and skills transferred to students in each school.

Meanwhile, during the Japanese occupation (1942–1945), Indonesian education underwent drastic changes from caste-based schools to schools which were intended for all residents. The Japanese military government in Indonesia abolished European-nuanced schools and turned them into public schools with a curriculum that was adopted from the Japanese education system. The Japanese military also prohibited the Western language and altered the language of instruction into two main languages, Malay and Japanese. During the Japanese occupation, education in Indonesia underwent major changes.

After its independence in 1945, Indonesia again adopted the Dutch education system. The leaders in the early days of independence were those who experienced Dutch education and they considered education in Indonesia at the time of the Japanese occupation to have decreased in quality; thus, they decided to adopt the Dutch school curriculum at the beginning of independence.

Changes in the education system and its management occurred when Indonesia entered the decentralisation era (post New Order, 1998), during which Indonesia's educational system received a lot of influence from Australia and the United States. School-based management that was adopted during that time led to the emergence of the curriculum based on Education Unit Level (*Kurikulum Tingkat Satuan Pendidikan*), or curriculum that gave schools the freedom and space to develop school-based, or area-based, curricula. With this curriculum, each region determined its local content; for instance, the Province of Central Java determined that Javanese Language would be a mandatory local content at the provincial level, while the Solo City government established that the subject of Batik (a traditional clothes of Javanese people) would be the local content at the city level.

After the school-based curriculum era, Indonesia then implemented a competency-based curriculum (*Kurikulum Berbasis Kompetensi*), and finally in 2013, the country implemented a curriculum known as the 2013 Curriculum that featured a scientific approach as a main issue in science. In the current curriculum, several new pedagogical approaches are introduced, such as thematic integration at the elementary schools and integrated science as an approach to learn science in junior high schools.

What we need to discuss further at this point is the way in which cultural pluralism is applied in science learning. The discussion focuses on current conditions, where all schools in Indonesia have implemented the 2013 Curriculum which was recently revised in 2017. The revision in 2017 mainly emphasised the scientific approach, focusing on learning of science and technology as well as strengthening scientific literacy training, numeracy literacy, and digital literacy.

The discussion utilises the two approaches proposed by McFarland, that is, multicultural education and critical pedagogy. The discussion is directed towards answering the questions: Have the curriculum and education system in Indonesia adopted multicultural education and critical pedagogy? Have the teachers paid attention to the issue of cultural pluralism and adopted it in their science classes?

Adaptation of Cultural Pluralism in Indonesia

In the 2003 Law on National Education System, it was guaranteed that every citizen has the right to education and education in Indonesia must be carried out in a democratic and non-discriminatory manner by upholding human rights, religious values, cultural values, and national pluralism. This statement is outlined in Article 4 Number 1 of the Law.

According to Tilaar (2004), multicultural education is education that prioritises the same rights, including access to education. In the discourse of multiculturalism, the process of democratisation is a necessity to recognise the rights of every human being regardless of skin colour, religion, cultural background, or gender.

The issue of multiculturalism reappeared in Indonesia in the era of decentralisation when government policies began to change perspectives about politics, economy, and culture, including centralised education for all regions. Decentralisation was adopted in 2000, where its impact encouraged the emergence of regional spirit in all aspects of society and nationality, including in education (Nurcahyono, 2018). As described earlier, this change is marked by the implementation of the Education Unit Level Curriculum that gives autonomy to regions and schools to provide local learning content in their schools.

Based on the opinion of Banks (2018, 1993, 1976), multicultural education can lead to opportunities for individual students to understand who they are by using the perspective of other cultures, recognising that there are various cultures around them, reducing the occurrence of discrimination against race, religion, colour, and culture, where the latter would provide opportunity for all students to master basic knowledge and skills, that is, reading, writing, arithmetic, and communicating.

The benefits and goals presented by Banks as mentioned earlier leads us to the discussion on critical pedagogy which is related to how teachers in schools are able to better teach multicultural students by applying the principle of equity. Under the principle of equity, students are assumed to learn better and faster if they are facilitated according to their prior knowledge or prior conditions. This includes the adoption of contextual learning methods, which is argued as the easier way for students to understand the concepts learnt because the approaches or examples used by the teacher in the classroom come from what is recognised by the students from their daily life in the society.

Historically, the origin of multiculturalism education in Indonesia can be traced back to the introduction of the slogan *Bhinneka Tunggal Ika* (Unity in

Diversity) which was a motto used by the Majapahit Kingdom as a cultural policy in maintaining harmony in the religious life of the people in the kingdom (Nurcahyono, 2018). However, this slogan also served as a legitimacy for the state to unify education policies during the New Order era (Suharto Period, 1966–1998) as it was viewed as another form of nationalism. During that period, education was no longer caste-oriented as practised during the Dutch colonial period. Arguably, access to education in the country has become increasingly open to people of all walks of life; however, the education system and specially applied science learning have yet to accommodate regional uniqueness which is reflected in the habit and lifestyle of the indigenous ethnic groups in each region.

To examine whether regional specificities have been included in science educa-tion in the Indonesian education system, a series of questions may be asked as follows: (1) Has content related to the region been included as material that can be learned in science subject? (2) Has local knowledge been offered by the teacher as one of the knowledge areas that should be considered as an alternative to solving science problems? (3) Has local technology been introduced to the students as a tool for solving science problems?

Research on how indigenous knowledge was described and taught in biology textbooks (life sciences) from 1951 to 2012 showed that the biology textbooks for high school in the 1950s and 1960s contained more elements of indigenous knowledge than the textbooks used in the 1970s, 1980s, 1990s, and 2000s. Indigenous knowledge on the topic of animals was found with high frequency in the textbooks published in the 1950s, while in the 1960s, it was indigenous knowledge on plants, and in the 2000s, it was on the topic of biodiversity and ecology (Ramli, 2009).

In this regard, the elements of multicultural education that are raised concern only the introduction of scientific facts but have yet to cover local knowledge and technology that can be used to solve problems. One example that surfaced in the high school biology textbooks of the 1970s was the *Wawo* (*Lysidice oele*) and *Palalo* (*Palalo viridis* or *Eunice viridis*) worms, which are highly nutritious and often consumed by the people of Sumba in East Nusa Tenggara. The harvest season for these worms, which is only once a year, is a tradition known as the *Bau Nyale* tradition. The knowledge about this worm in the old biology textbook was presented in the chapter of kingdom animalia on phylum of annelida. Unfortunately, the current biology textbooks in Indonesia no longer reveal these worms as annelida cases that high school students need to study.

What Do the Teachers Think about Cultural Pluralism?

When teachers in Indonesia were asked about Indonesia's diversity, they showed awareness of this issue. This is not surprising as the issue has been strongly disseminated through the mandatory training programmes conducted for civil servant teachers. The teachers' understanding may not be detailed when asked about certain ethnic rituals, but in general, they understand that Indonesia is an ethnically and culturally diverse country.

The national insight seems to be getting stronger again when the current president of the Republic of Indonesia, Joko Widodo, often asked students, teachers, farmers, fishermen, villagers, and even bureaucrats of their knowledge regarding the tribes in Indonesia, the name of the province, the names of the fish and others related to the uniqueness of the region as well as the issue of pluralism in the country during his field visit to the regions, specially during audiences with the community. Even though this is just considered a typical way of President Joko Widodo interacting and mingling with the people, it has raised the enthusiasm of the people, including students and even teachers, to hone their knowledge of Indonesia again. Moreover, news about President Jokowi's actions is often officially broadcasted live and replayed by the Presidential Secretariat through the YouTube channel; it is also repeatedly broadcasted on the national TV channels and radio broadcasts and published in newspapers both online and offline, to the point that the news quickly spreads to the public.

What Do the Teachers Think about CRSP?

Culturally Relevant Science Pedagogy (CRSP) or Culturally Relevant Science Teaching (CRST) is a pedagogy or teaching that can actively involve students from various backgrounds to enjoy science learning equally. CRSP can be said to be the embodiment of the concept of critical pedagogy.

To find out how teachers adopt CRSP, we conducted an online survey focusing on the topic of ecosystems and environmental pollution. The respondents of the survey were elementary school, junior high school, and high school teachers ($N = 212$). The survey asked about the content, learning strategies, and whether the teachers have applied the principle of equity in learning or otherwise. Most of the survey respondents were female (67.9%). The respondents were of various age groups; however, teachers of the age between 50 and 60 years (32.5%) made up the largest group, with many of the respondents having worked as teachers for between 11 and 20 years (41%). About 59% of the respondents were junior high school science teachers. As many as 35% of the teachers taught in schools that are in sub-districts, 27.4% in small cities, 25% in villages, and 12.3% in big cities. The respondents were of relatively diverse ethnicity; however, as much as 88.2% are Javanese.

In response to the question whether the respondents included local contextual elements in the learning materials for ecosystems and environmental pollution in the classroom, 46.2% of the teachers stated that they rarely do so, 15.6% reported never doing so, and only 4.2% reported having always integrated contextual cases in their learning. When asked whether they included images and local cases as learning resources, only between 41% and 49% of the teachers reported doing so.

What is interesting is that around 49.1% of the teachers prioritised using materials from the textbooks, and if we cross check the content of the materials in the textbooks, what is conveyed are national cases, for example, illegal logging, air pollution because of burning of trees in the forest, and cases of Lapindo mud pollution which became a national disaster in Sidoarjo, East Java. The textbooks used in Indonesia are almost the same as they are published by Erlangga Publisher. The government also facilitates schools in the regions with digital textbooks that can be accessed at https://bse.belajar.kemdikbud.go.id.

Respondents were also asked of their effort to understand their students, especially those of diverse cultural, economic, and environmental backgrounds. As many as 35.4% of the respondents said they rarely did it, and 34.4% said they often did. Likewise, 43.9% of the teachers often asked their students the local terms (in the local language), including asking the students about pollution in their place of residence. In relation to providing students the opportunity to submit their opinions freely, about 49.5% of teachers said they often did it, while 17.9% of respondents said they always gave the students the opportunity to do so.

Concerning grouping of students for assignments or learning activities, about 10.8% of the teachers reported that they always performed a variety of groupings based on cultural background differences of the students; however, 36.3% answered that they had never considered grouping students based on their cultural diversity. A possible explanation for the response could be attributed to the fact that there are several schools where the students come from only one ethnicity, which thus results in such a response from the teachers. When asked whether the teachers gave the opportunity for students from minority or marginalised groups (female students/students from villages/students of weak economic backgrounds/students of ethnic minorities) to lead discussion groups, as many as 34% of the respondents said that they rarely did this, and only 3.3% said they always did, while 21.2 % of respondents reported never doing so.

We also asked respondents of their perceptions about equity and equality and whether they have practised it in the classroom, and how it is implemented in their teaching or in the classroom. As many as 39% of the respondents admitted that they could distinguish between equality and equity, 49.6% were doubtful, and 10.8% reported that they are not able to do so. To confirm their responses, we presented an illustration of equity and equality and then asked the teachers which picture was often applied in their science class. As many as 53.3% of teachers stated that equity is the most important thing to be considered in the science class, while 13.2% stated that equality should be applied. Meanwhile, 28.3% of respondents stated that both must be applied, and around 4.2% stated that they did not know which one should be implemented.

It is also interesting to note that most of the teachers chose to give students the same time, opportunity, atmosphere, learning resources and learning media (59.4%) which suggests that they prioritise equality rather than equity; only 13.2% of the teachers facilitated all of these differently. In addition, 13.7% of the teachers questioned and checked students' needs and provided what was needed, while 10.8% of the teachers gave freedom to the students to determine their own media, learning resources, and learning styles.

Why do science teachers in Indonesia tend to provide the same facilitation to all students, and do not show great concern about cultural pluralism? The reason for this is purportedly because of the science curriculum that is applied in schools which is still considered too dense and takes a lot of time to complete. The teachers also assumed that if they do not complete all the materials, then the students would have difficulty when sitting for the national exam.

The national exam is a system of graduation selection for elementary, junior high, and high school students in Indonesia before it was abolished in 2020. The national exam for high school students also determines whether they can take part in the National Higher Education Admission Selection. Several research findings seem to show that teachers are very burdened by the existence of the national exams and for this reason, they cannot freely arrange science learning strategies and can only concentrate on training their students to be successful in the national exams.

Although the results of the survey cannot represent all categories of teachers in Indonesia, it can be depicted that science teachers in Indonesia do not actually understand the principles of CRSP well and do not apply all the principles in their teaching. Most of them do not really care about the differences in their students' cultural background or cultural pluralism that exists in the classroom. It is suspected that the teachers may think that this issue is a common thing, or the teachers may not feel the impact of students' learning differences attached or linked to the students' cultural differences.

However, the effort of several teachers who made the attempt to check their students' prior knowledge and cultural backgrounds, group the students based on their differences in competence, and then introduce contextual problems and local cases that are familiar to the students should be appreciated as efforts to implement CRSP in the classroom.

The condition of science teachers in Indonesia is exactly like the situation described by Gloria Ladson-Billings in chapter "'Yes, But How Do We Do It?' Practising Culturally Relevant Pedagogy", in *White Teachers/Diverse Classrooms*. She stated that teachers in schools are basically expected to follow the prescribed curriculum that has been established by the state and the local education council who approved it, and in most schools, the prescribed curriculum means learning based on the textbooks (Ladson-Billings, 2006).

Ladson-Billings clarified that culturally relevant pedagogy can be explained by three components that must be mastered by teachers to be able to help the success of their students, namely academic achievement, cultural competence, and socio-political consciousness. Academic achievement is the main function of educating children in school. Teachers who agree with this understanding prioritise empowering students' thinking and providing support for students' intellectual lives. The teachers in this group, however, are not aiming at anything other than cognitive achievement. Teachers who understand the relationship between culture and pedagogy are always thinking deeply about what they should teach to their students, and always perform self-reflection, asking why their students should study certain materials outlined in the curriculum. These teachers are usually not tied to the textbooks but use all forms of learning resources that students are familiar with, for example, videos, films, music, folk tales, or students' funds of knowledge.

Cultural competence is a competency that can help students recognise and be proud of their cultural background, in the form of beliefs, perspectives, and real actions, when students access a broad culture, which could be a place where they get the opportunity to improve their socio-economic status and additionally, make decisions about the future they want. In other words, teachers who have cultural competences are teachers who can focus on improving the lives of their students, their families, and the communities where they live.

One of the facts that can be revealed is the experiences of high school graduates from the Papua region, which is a part of eastern Indonesia, when they get the opportunity to study at universities in Java Island, which is the most populous island in Indonesia, and the most developed island in the country. There are many culture shocks that these students face and experience, particularly the way of life and utilisation of technology. The Javanese, who make up most of the population on Java Island, are known to be diligent workers. Some assignments given by the lecturers are mostly IT-based, which the students from Papua are not familiar with. It may be that the Papuan students are not given a lot of exposure and experience of using information technology when learning at school in their area because of the limited facilities and access to information technology. Additionally, the style of communication also differs, and this requires the Papuan students to understand the Javanese style of communication that dominates the universities in Java.

Another interesting experience that we observed was related to science learning in the Karimunjawa Islands, which are in the Jepara Regency, Central Java Province. This archipelago consists of small islands inhabited by approximately nine ethnic groups, with the dominance of Javanese and Buginese tribes. These two ethnic groups have opposite mindsets. The Javanese are known for their non-candid speech and tend to choose a way of arguing that prioritises explanation (reasoning-evidence) before stating claims. Meanwhile, the opposite is applied by the Buginese and most of the other tribes in East Indonesia. As the school is in Central Java, Javanese is the local content that must be taken by the elementary school students in Karimunjawa irrespective of their ethnic group even though the teachers are aware that it is very difficult for children from non-Javanese tribes to learn Javanese as they still use the local language of their tribe in their daily living with the family and in their community's environment. Carving is a local content subject in Jepara Regency which is also compulsory in several schools although it is known that the Buginese people are known as skilled sailors and ship and boat builders. For the children of the Buginese tribe, this could have a negative effect because they cannot learn the knowledge of their ancestors in school. Apart from the Buginese people, the Bajo people are also boat people as most of their life is spent on boats. Similarly, the Bajo children are also studying the dominant Javanese culture in school, and this is considered by the Bajo tribe as an effort to change the future of their children to be better economically and to no longer live on boats.

Socio-political consciousness is the teacher's awareness of socio-political issues locally (in the region) and in the school environment and bringing understanding of these issues into classroom learning. An experience was narrated by a junior high school teacher we interviewed regarding her teaching of ecosystem and environmental pollution in Solo. This teacher brought up the case of the *Bengawan Solo* River pollution, one of the largest and longest rivers on the island of Java which passes through the city of Solo. She introduced the changes that occurred in *Bengawan Solo* by inviting students to listen to and appreciate the lyrics of the song *Bengawan Solo*, written by Gesang in 1940. In the song, Gesang describes the condition of *Bengawan Solo* during that era, where the river was the main mode of transportation connecting cities on the island of Java, and clean water was abundant and flowing swiftly through the river. To compare the condition of *Bengawan Solo* in the past and the present time and what has happened to the river, the teacher brought the students to visit *Bengawan Solo*, carry out observations, and check the water samples so that the students could understand why *Bengawan Solo* could no longer be passed by boats and the water was no longer clear.

Conclusion

Cultural pluralism in Indonesia is a gift that science teachers must interpret and translate into science learning strategies in schools. Teachers have strong awareness of cultural pluralism, but the knowledge and skills to use it in pedagogy of science learning are still limited. Therefore, efforts are needed to help teachers realise and apply it at a practical level.

Even though the 2013 curriculum has been revised so that it can accommodate current changes and developments, it must at the same time continue to be able to focus on pluralism accommodation policies in the regions; guarantee the freedom of teachers to improvise, innovate, and create to adapt the context and content of science learning to real conditions in the regions; and pay attention to the different cultural background aspects of the students. Therefore, while the curriculum should set competency standards, clear learning progressions and core ideas, it should emphasise upon teachers to always take advantage of local learning sources and determine learning strategies based on student pluralism.

The science learning approach in the curriculum, which was originally monodisciplinary, must be changed to be interdisciplinary, even transdisciplinary. This approach needs to be applied because in accordance with the CRSP concept, teachers must bring contextual or local problems to their students. Additionally, to be able to solve these problems, an interdisciplinary or even transdisciplinary approach is required as mono discipline is deemed incapable of solving these problems. Students should be equipped with multidisciplinary insights that would enable them to review problems in detail and propose accurate problem solutions. Approaches such as science, technology, engineering, and mathematics (STEM) need to be accommodated in the science curriculum.

Efforts to integrate learning themes have been described in the 2013 Curriculum, in the form of integrated thematic learning in elementary schools and the integrated science approach in junior high schools. Integrative thematic learning is the implication of multidisciplinary-based learning, where students will learn based on certain themes, which involves several subjects, such as Indonesian language, social sciences, and natural sciences. However, teachers' expertise in applying thematic integrative learning is still very limited, such that in some schools, teachers teach mathematics and science separately. The integrated science approach has also not been fully implemented in schools because many teachers do not yet understand how to integrate interdisciplinary concepts in science.

Apart from the curriculum, another policy considered to be distressing by teachers and students is the national exams which serve as a determinant of student graduation and even student success in the future. In 2020, the Ministry of Education and Culture of the Republic of Indonesia finally abolished the national exams. Even though it is accompanied by pros and cons, this policy is expected to give freedom to teachers to break down the curriculum to be more flexible and ultimately accommodate pluralism in schools.

Other policies that need to be pursued are strengthening educational research and science learning that accommodate the CRSP components, from both among teachers as practitioners in schools and among researchers at the higher education level. Research on CRSP in Indonesia is still relatively limited. Some of the ones already available can be assumed to be related to CRSP, for example, research on exploring the local issue as a source of science learning; however, it is still rare. Research links and matches between schools and universities need to be sharpened and strengthened so that the direction of research in tertiary institutions can be applied and implemented in schools.

Training for teachers to promote CRSP can be carried out in co-operation with regional universities that have teacher education institutions. Utilising the learning community of teachers as training partners can be selected to improve the quality of CRSP-based learning in Indonesia.

References

Azra, A. (2010). Cultural pluralism in Indonesia: Continuous reinventing of Indonesian Islam in local, national, and global contexts. *Annual Conference on Islamic Studies, 10*(November), 4–13.

Banks, J. A. (1976). Cultural pluralism and contemporary schools. *Equity and Excellence in Education, 14*(1), 32–36. https://doi.org/10.1080/0020486760140110

Banks, J. A. (1993). Chapter 1: Multicultural education: Historical development, dimensions, and practice. *Review of Research in Education, 19*(1993), 3–49. https://doi.org/10.3102/0091732X019001003

Banks, J. A. (2018). *An introduction to multicultural education.* Pearson Education.

Bernstein, R. J. (2015). Cultural pluralism. *Philosophy and Social Criticism, 41*(4–5), 347–356. https://doi.org/10.1177/0191453714564855

Ladson-Billings, G. (2006). "Yes, but how do we do it?" Practicing culturally relevant pedagogy. *White Teachers/Diverse Classrooms.* http://fordhamatsdc.files.wordpress.com/2011/08/ladson-billings_g-_yes_but_how_do_we_do_it.pdf

McFarland, K. P. (1999). *Cultural pluralism: The search for a theoretical framework.* https://files.eric.ed.gov/fulltext/ED434870.pdf

Nurcahyono, O. H. (2018). Pendidikan multikultural di Indonesia: Analisis sinkronis dan diakronis (Multicultural education in Indonesia: Synchronous and diachronic analysis). *Habitus: Jurnal Pendidikan, Sosiologi, & Antropologi, 2*(1), 105. https://doi.org/10.20961/habitus.v2i1.20404

Pantoja, A., Perry, W., & Blourock, B. (1976). Towards the development of theory: Cultural pluralism redefined. *The Journal of Sociology & Social Welfare, 4*(1), 125–146. https://scholarworks.wmich.edu/jssw/vol4/iss1/11

Ramli, M. (2009). *Analisis substansi pendidikan multikultural Sains di buku pelajaran Biologi untuk SMA (The analysis of multicultural science education substantives in the high school biology textbooks).* In Proceeding Biology Education Conference of Seminar Nasional IX Pendidikan Biologi FKIP Universitas Sebelas Maret, Indonesia NS (pp. 135–141). https://jurnal.uns.ac.id/prosbi/article/view/7430/6598

Tilaar, H. A. R. (2004). *Multikulturalisme: Tantangan-tantangan Global Masa Depan dalam Transformasi Pendidikan Nasional (The future global challenges in national education transformation).* Grasindo.

7 Case Study in Malaysia

Is Culturally Responsive Pedagogy Possible?

Siti Nur Diyana Mahmud

Introduction

Malaysian school classrooms are characterised by diversity where students come from various cultural, racial, socio-economic and different cognitive ability backgrounds. As more students from diverse backgrounds converge in the classroom, efforts to identify effective methods to teach these students mount, and the need for pedagogical approaches that are culturally responsive intensifies. It is important for the teachers to understand, acknowledge and address these diversities when planning lessons and conducting teaching and learning in the classroom. Today's classroom requires teachers to educate students who vary in culture, language, abilities and many other characteristics. To meet this challenge, teachers must employ not only theoretically sound pedagogies which are based on learning theories but also culturally responsive pedagogy (CRP). Teachers must create a classroom culture where all students regardless of their cultural, linguistic, and socio-economic background as well as ability are welcomed, supported and provided with the best opportunity to learn. Addressing diversities in the classroom is important to create an all-inclusive atmosphere and ensure equity is achieved.

CRP facilitates and supports the achievement of all students. In a culturally responsive classroom, effective teaching and learning occur in a culturally supported, student-centred environment, where the students' strength is identified, nurtured and utilised to engage students in learning and promote their achievement. For some students, types of discourse in the classroom contrast with the cultural and linguistic practices at home. To engage students in learning, it is important that teachers facilitate students to bridge this discontinuity between home and school (Allen & Boykin, 1992). Furthermore, a culturally responsive teaching and learning environment minimises the students' alienation as they attempt to adjust to the different "world" of school (Ladson-Billings,1994). This chapter defines CRP and explains how it might be used effectively to address the instructional needs of a diverse student population.

DOI: 10.4324/9781003168706-9

Culturally Responsive Science Pedagogy

CRP emphasises equity and inclusivity where students' cultural capital and funds of knowledge in their learning are acknowledged and affirmed. Funds of knowledge are defined as the bodies of knowledge, including information, skills, strategies, ways of thinking and learning, which underlie household functioning, development and wellbeing (González et al., 1993). Examples include parents' occupation, house chores, home language, holiday and tradition. CRP utilises student's funds of knowledge as a prime learning resource to promote high academic performance, (inter)cultural competence and critical cultural consciousness (Gay, 2018). From the Vygotskian perspective, cultural tools and activities mediate learning and teaching. Vygotsky (1978) argued that learning should be authentic and relevant to the student's daily life and practices in a community or culture. In elementary education, these tools include, but are not limited to, physical aspects of the environment such as play equipment and resources.

A focus on funds of knowledge provides a framework to recognise how children's knowledge and interests arise in, and are stimulated by, the contexts of their intent participation in everyday activities and experiences. Moreover, their funds of knowledge are supported by the social relationships that are integral to developing knowledge of the world. Moll et al. (1992) and Shulman (1987) believed that if teachers established understanding of local funds of knowledge as a form of teachers' professional knowledge, this could inform curriculum through teaching and learning that are organised around children's interests and questions, as well as build respect for diverse communities, thereby improving children's educational experiences and outcomes.

Reciprocal and Responsive Relationship for CRSP

CRP involves reciprocity, understanding and respect of differences in teaching. CRP follows the constructivist approach which is learner-centred with a balanced agency in teacher–student relationship. Emphasis in teaching and learning is given to metacognitive inquiry and higher-order thinking skills. The teacher scaffolds students learning where knowledge and learning practices are expressed in different cultural contexts and shared in heterogeneous learning communities through self-reflectivity (Perso, 2012).

CRP engages students in learning and impacts their performance through a balanced agency in student–teacher relationship based on collaborative and peer learning. Balanced agency between students and teacher in teaching and learning can be created and achieved through intersubjectivity. Intersubjectivity involves developing shared purpose and meaning between the teacher and students in learning and teaching experiences. Intersubjectivity aligns with the inquiry approach and is important in building the curriculum from the student's authentic interests. Creating intersubjectivity requires the students and teacher to contribute to the learning exchange where the contribution may not necessarily be equal.

Furthermore, students may create ideas with the teacher, and the intent is for the teacher to simply listen and reflect on the student's thinking and understanding. Meaningful student engagement in learning depends on the teachers listening closely to the students. Clark et al. (2005) argued that listening to students is related to power relationships in teaching and learning. Usually, by listening to the students, teachers could identify students' interests and funds of knowledge that arise from these conversations.

Moreover, in a culturally responsive classroom, learning and knowledge construction is situated, where student learning is mediated by social interactions, activities, cultural tools and practices. Term *Distributed cognition* describes the shared processes by which this learning occurs and is mediated. Considerations of reciprocal and responsive relationships and distributed cognition support the concept of learning as a process of transformation of participation employing cultural tools (Gutiérrez & Rogoff, 2003).

Culturally Responsive Instruction

Culturally responsive instruction focuses on both high learning expectations and academic rigour and scaffolded learning activities that utilise students' funds of knowledge in teaching and learning. Students of diverse sociocultural backgrounds perform better when teachers have high academic expectations for them. Furthermore, students perform better when teaching is authentic and relevant to students' daily life and background. Culturally responsive instruction enables students to see themselves as the main actors in their learning process (Kalantzis & Cope, 2008).

Seven common characteristics of CRP encompass teachers' ability to (i) create a caring, respectful classroom climate that values students' cultures; (ii) build connection between academic learning and students' prior cultural and language knowledge; (iii) make instruction meaningful and relevant to students' lives; (iv) fully integrate into the curriculum local knowledge, language, and culture; (v) hold high academic expectations for all students; (vi) create a more collaborative and challenging learning environment away from traditional teaching practices (memorisation and lecturing); (and vii) build trust and partnerships with families, especially marginalised ones (Perso, 2012).

Culturally responsive feedback occurs when teachers provide immediate, critical and ongoing feedback in well-designed activities. Research also shows that responsive feedback has a positive impact on low achieving students (Fuchs & Vaughn, 2012). Overall, students' intrinsic motivation could be enhanced through recursive feedback, co-produced learning and respect of cultural differences replacing the metaphor of extrinsic reinforcement.

Instructional scaffolding is another essential element of a culturally responsive approach. Teachers' ability for pedagogical scaffolding when designing learning repertoires could bridge what students already know (prior knowledge) and are familiar with to the intended learning (new learning). Scaffolding may include an epistemic framework of mixing different multimodal activities (experiential, conceptual, analytical and application) (Kalantzis & Cope, 2008) to enhance deeper understanding. Thus, teachers' ability to act as co-designers of materials is critical. Teachers and students co-create the knowledge in the classroom and are not merely utilising nationally selected materials in the textbooks, which are usually ethnocentric. Researchers have argued that diversity should be present in materials to reflect students' cultural backgrounds (Gay, 2018; Ladson-Billings, 2014).

In science education, a number of studies have investigated CRSP in the current practices of science teaching and learning. The context of these studies varies from indigenous students to migrants, as well as non-Western context. For example, the study by Glynn et al. (2010) demonstrated how the science teachers built trusting and respectful relationships with their Māori students by facilitating connections between Western and Māori worldviews of science. They shared their teaching role with Māori elders (kaumātua) and members of the extended family of their students (whānau). The teachers learned a great deal from their Māori students who became highly engaged and agentic in their science learning. The students took collaborative responsibility for asking learning questions and sought information on science topics from both the Western and Māori worldviews.

In the past decade, lines of research in science education have focused on the development of CRP, which is in essence, situated science learning that embraces local culture and knowledge that are contextualised in non-Western axiology. There have been considerable debates concerning the way the school curricula, and in particular, the science curricula, should be framed in order to respond to the cultural diversities encountered in nations especially those that are non-Western such as in Malaysia.

Science Curriculum in Malaysia

Science curriculum development in Malaysia has been driven by the goal to enhance learning and is related to national development goals as well. The curriculum has an important influence on students' science learning. Science learning is influenced by various factors, including a robust curriculum, students' readiness to learn and teachers' willingness to guide students. Students' readiness to learn depends on the students' ability to adapt the initial concepts of science they have with the concepts of science taught in school as well as by using appropriate learning techniques to obtain excellent results. For this reason, teachers need to understand students' initial perceptions of a science concept to be taught.

The changes made in the science curriculum are not only in terms of content but also in terms of teaching methods. Teachers had to make some adjustments, namely in terms of content and teaching methods and modify activities and new ideas taken from the West to suit the local environment.

The changes in the Primary School Science Curriculum were made to meet the current needs based on the latest developments and the requirements of the National Education Philosophy, which is highly concerned with the physical, emotional, spiritual and intellectual development of the individual students. These changes are based on students' knowledge and understanding of science in terms of their skills and attitudes towards the science subject itself. Furthermore, these changes in the science curriculum were also driven by shortcomings in the previous curriculum. For example, in special science projects, science teaching was based on two priorities, namely examinations and textbooks. Teaching only revolved around the mastery of facts. It emphasised the mastery of scientific facts with little application of social and religious values. Meanwhile, the topic of 'Nature and Man' was able to develop students' knowledge about humans, the environment, society, and the interaction between them. However, there was a lack of competence among teachers in integrating the subject content and in using inquiry in teaching. In addition, the subject content was not in-depth, and students were weak in science subjects in high school. Although the implementation of KBSR showed success in fostering 3M (reading, writing and counting), the integration of knowledge and values did not occur in the classroom because the focus was on achieving good grades in examinations.

KSSM was introduced because there is a need to make the National Curriculum more holistic, relevant and in line with the needs of the twenty-first century. In the twenty-first century, the country needs human capital who are critical, creative and innovative in order to contribute to the country's development and to have global competitiveness.

Although the science education curriculum has undergone various transformations since its first introduction during the colonial times up to the present, in this highly centralised system, teachers have always concentrated on teaching students to perform very well in the external examination conducted at the end of the year which has been used as a measure for reporting school performances. Since the performance in the examinations is what counts, parents have always pressured the schools and teachers to teach to the examination curriculum. Thus, a teacher as a researcher is not considered an important ideal that Malaysian school teachers should have.

Science Teacher Education in Malaysia

The Malaysian education system is highly centralised. As a result, the development of science teacher education is closely related to the development of the national education system. The structure of teacher education programmes, that is, pre-service and in-service trainings, is also developed based on the need of the educational system, socio-economic and politics of the country as well as the impact of globalisation.

In general, science teacher education in Malaysia began with the training of British educators as their science curriculum was adapted for use in Malaysia after achieving its independence. In the 1980s, when the medium of instruction was changed to Malay, the Faculty of Education at Universiti Kebangsaan Malaysia (UKM) was the first faculty responsible for training science teachers to teach in Malay. Another development in training for science teachers is the policy of science and mathematics teaching in English which was established by the Ministry of Education in 2003. Nowadays, science teachers who teach in primary and secondary schools are mostly graduate teachers. Primary school science teachers are trained by the Institute of Teacher Education as general science teachers. Meanwhile, high school science teachers are trained by the respective universities based on educational standards set by the Malaysian Qualifications Agency. For in-service training, the main purpose is to motivate teachers for any changes in the science curriculum that have been implemented. At the same time, in-service courses also include science teachers pursuing advanced study programmes such as postgraduate studies.

Although training for science teachers in Malaysia has gone through some developments, issues related to the quality of teachers are still raised. Weaknesses in teaching strategies reported in the 1990s by the Federal School Inspectorate are still repeated in the new century, especially teacher-centred teaching, 'chalk and talk'-patterned teaching, teaching that does not practice the application of noble values, and examination-oriented teaching. It was found that teachers use teaching strategies that are more inclined and oriented towards examination (especially behaviourism approach and drills that focus on memorisation) compared to teaching that emphasises thinking skills (Phang et al., 2014).

In general, most studies have found that there are significant differences in the level of Pedagogical Content Knowledge (PCK) of new teachers compared to experienced teachers. The level of PCK among novice and pre-service teachers is highly dependent on algorithms and memorisation of formulas, tips and rules, and shows inability to provide reasons based on conceptual knowledge as well as inability to solve problems that require "two stages or processes" of solution. Most of the experienced teachers exhibited content-centred and comprehension-centred PCK as well as student-centred learning activities. Several studies have found that there are misconceptions of science concept among teachers, especially among novice teachers and pre-service teachers. These shortcomings have caused students to experience learning difficulties in science subject.

The Study Context

This study utilised a qualitative approach which focuses on how people interpret their experiences and the meaning they attribute to their experiences. The purpose of this study was to describe how the teachers in the case study build meaningful bridges between students' home experience and scientific concept and to identify the challenges in implementing culturally responsive teaching. Such a purpose suggests a case study design since: (a) the focus of the study is to answer "how" or "why" questions; (b) the behaviour and interpretations of those involved in the study cannot be manipulated; and (c) the contextual conditions are relevant to the phenomenon under study. For these reasons, the case study design was chosen as the research design of this study.

The research participants consisted of four elementary novice science teachers who taught in urban schools. Each of the science teachers for this study was purposefully selected for their willingness to prepare the CRSP lesson plan together with the researchers. The participants' recruitment was through snow-ball sampling technique. The research participants selected for this study were required to develop CRSP lesson plans with the researchers and implement the lesson plan in their science learning class. Each of the participants was given a pseudonym, namely T1, T2, T3 and T4, respectively to protect the confidentiality of the participants. Participants T1 and T3 were male teachers, while T2 and T4 were female teachers. All the research participants have less than five years of teaching experience.

The interpretivist approach was used to interpret and understand the teachers' perception of CRP in Science teaching and learning. It is the preferred research paradigm because this study focused on different perceptions held by teachers and students. The focus on alternative meanings within the sample led to the selection of interpretivism as the relevant paradigm. It is crucial for an inter-pretivist researcher to understand motives, meanings, reasons and other subjective experiences which are time- and context-bound.

In this study, classroom observations were conducted in a structured way by employing an observation schedule that consisted of forms prepared prior to the data collection, where the behaviour and situational features to be observed and recorded during the observations are outlined. In this study, the classroom observations were conducted four times by the researchers. The observation form included students' position in the classroom, classroom environment, description of instructional methods, and description of students' behaviour in the classroom. The student behaviour observation checklist included students' listening, writing, reading interaction with the teacher and interaction with other students. These observation forms were used to ensure that collected data were guided by the research questions of the study. The observation forms also functioned as flexible guidelines for data collection by listing topics of interest and providing space to record notes about new themes that emerged during observation. The classroom observation provided the researcher a deeper under-standing of the interactions taking place between the teacher and the students in the real classroom setting.

Findings and Discussion

The findings are discussed according to the research questions. The first research question was to describe how the teachers in the case study build meaningful bridges between students' home experience and scientific concept while the second research question sought to identify the challenges in implementing culturally responsive teaching.

How Does the Teacher Build Meaningful Bridges between Students' Home Experience and Scientific Concept?

The data from the observation showed the teachers' responsive interactivity with the students during the lesson helped to build a meaningful connection between the lesson and students' daily life. The responsive interactivity can be observed several times during the observations. For example, at the beginning of the lesson, one of the teachers used the example of fox and chicken to explain the concept of prey and predator. This example was planned by the teacher in the lesson plan. Nevertheless, the students who live in urban areas were not familiar with these animals. This confusion was explicitly exhibited by the students during that lesson. The teacher was responsive to this situation and deviated from the example stated in the lesson plan by using the funds of knowledge from popular culture which was familiar to the students. Researchers mentioned that students' and teachers' funds of knowledge may sometimes conflict with each other (Maitra, 2017). Therefore, it becomes even more important for teachers to understand their students. From a sociocultural perspective of pedagogy as "responsive interactivity" (Edwards, 2001), responsive teaching comes from teachers and children knowing each other well and sharing purposeful learning in positive, reciprocal interactions that have a joint focus, hence the notion of co-constructing the curriculum.

Furthermore, in the aforementioned example, the teacher built the connection between the students' funds of knowledge and the science lesson through a reciprocal relationship. Reciprocal and responsive relationships imply that inter-subjectivity and the opportunity for sustained interactions and dialogue are important in developing meaningful learning. The range of funds of knowledge, including students' home language, daily experiences and family background is essential to support the reciprocal and responsive relationship. Jointly focusing attention on or initiating a conversation about something a student has shown interest in helps the student to make sense of the experience and build understanding of the topic of interest. For example, during the class observation, the students showed interest in the conversation about pets. The teacher utilised the conversation topic and linked it with the lesson which resulted in the students actively engaging in the learning and being able to develop their understanding about the topic on that day. Taking this into consideration, intersubjectivity during the lesson largely depends on the students and teachers knowing each other well in order to be responsive. The basics of critical thinking and metacognition are also laid in such interactions and support students' integrated experience-based approaches that foster choice, inquiry and independence. Furthermore, there are implications for teachers' ability to engage in sustained interactions with students. Communicating with students to create intersubjectivity and act on their understandings and curriculum decision making may be influenced by teacher–student ratios or inter-pretations of teacher roles (Kontos & Wilcox-Herzog, 1997).

In addition to the responsive and reciprocal relationship between the teachers and students during the lesson, the teachers in this study encouraged the students to co-construct their understanding during the lesson through collaborative learning. Research indicates that direct and explicit collaborative learning improves student engagement and motivation, enhances problem solving and reciprocal peer learning. By co-constructing knowledge during the lesson, it enabled both the teacher and students to engage in collective learning where knowledge outcome is shared. For example, one of the teachers was observed to enable the process of co-constructing knowledge during the lesson by asking each group to share their drawing and understanding of the topic in front of the class. The first and second group who presented showed misconceptions about the lesson. The teacher then scaffolded the students' discussion concerning the first and second groups' answers and drawings. The third group managed to grasp the concept of the lesson and provided the correct answer. The first and second group then changed their answers and improved on their drawing after listening to the third group's presentation.

The Challenges in Implementing CRSP in Malaysia

This study identified the challenges in implementing CRP by analysing the teachers' reflection, the science curriculum, and science teacher education in Malaysia. Three themes emerged, namely: (i) lack of acquired skills, (ii) time constraint, and (iii) examination-oriented culture and system.

The teacher participants in this study were novice teachers who have less than five years of teaching experience. Their background might influence their perception of CRSP's implementation in the science lesson. All four participants agreed that implementing CRSP requires a set of skills involving identifying students' background and connecting the lesson with students' context. They expressed difficulty in performing both skills.

> Each student's experience is different and sometimes it is difficult to relate lessons to the student's background.
> Teachers require skill to make connections between the syllabus and the student's context. It will take more time to do that.
> By doing culturally responsive teaching, more unexpected questions and ideas will come from students. Teachers need to be more careful and skilled enough to focus on teaching, because sharing information from students are diverse and may deviate from the planned lesson.

These responses from the teacher participants concerning challenges in implementing CRSP signify a great shortcoming of teacher preparation programmes and professional development programmes in preparing the teachers for CRSP. Not enough is being done to extend ongoing support to teachers who are willing to implement CRSP. Perhaps a more sustainable, more collaborative methodology is needed to support the teachers' implementation of a theory into practice. Sleeter (2012) found that even when working with teachers who have already embraced the ideals of critical pedagogy, they ended up dismissing it because they did not know what to do with it in their classrooms. This study advocates a more hands-on, more praxis-oriented and more collaborative model of research design that calls for inquiry-based discourse and iterative action and reflection to further support the work of teachers.

Additionally, many researchers argued that graduates of teacher education programmes do not always transfer or apply the best practices they learn in their university courses to their teaching practice (Brown & Bentley, 2004; Gainsburg, 2012; Lloyd, 2013). Researchers attribute the lack of learning transfer to a variety of factors including factors related to course features (e.g. course length, if students had the opportunity to apply the tools learned in a variety of contexts during the course), the student (e.g. ability, personality, motivation) and the workplace (e.g. how much support is available, if there is an opportunity to use what was learned) (Gainsburg, 2012; Prebble et al., 2005; Van den Bossche & Segers, 2013).

Besides the challenges in terms of skill, the findings revealed challenges in terms of time constraints as a result of the requirement to complete the syllabus and the diverse student background in the classroom. According to the participants' reflection after conducting the CRSP lesson, CRP requires more time to be implemented in the classroom. The teacher participants reflected that they experienced insufficient class time.

> It will take longer for each lesson because each student has a different experience. Teacher needs to ensure the students in the classroom have the right and desired learning framework.
>
> The time required for a lesson is longer compared to normal class.

Furthermore, the teachers in this study stated their concern of not being able to complete the syllabus on time if they were to implement CRSP in their classroom. This is because, according to them, CRSP lessons will take a longer time compared to conventional teaching. This time constraint issue unfolds a bigger picture of this situation which involves the curriculum and the time allocated to complete it. This issue is also related to teachers' lack of autonomy in deciding the content and the way lessons are delivered. Although autonomy in the class is a very important feature and is even encouraged, many teachers do not show autonomy in terms of class work and only follow the designed syllabi. Some of the factors for this issue include the many boundaries and restrictions set up by schools for teachers to follow the syllabus, the large number of students in each class, the prearranged exams and time limitation, which altogether leave little or no space for teachers to design and apply their own tasks in the classroom.

Recommendation

To successfully implement CRSP in Malaysia requires systemic effort involving strengthening of the science curriculum and science teacher training. The following suggestions are a few recommendations in terms of curriculum; teacher training and collaboration; and teaching, learning and assessment to strengthen the quality of science curriculum and teacher training.

The first recommendation is to revise the science curriculum. There is a gap between the planned curriculum, implemented curriculum and achieved curriculum. Although the written curriculum developed by the Ministry of Education (MOE) specifically states that science education includes three main components which are knowledge, scientific skills and scientific attitudes, the taught curriculum does not reflect this focus very well and the examined curriculum does not reflect this balance either. Another critical component of the curriculum that needs to be addressed is the hidden curriculum where cultural elements practiced by the learning community significantly influence the teaching and learning processes. Furthermore, there is a need to involve professionals in science and the community to refine the science curriculum so that it is more contextual and based on everyday life. They can also be invited to school to teach students about some science concepts in one teaching session.

The second recommendation is in terms of teacher training and collaboration. The aspect of PCK as well as knowledge of context must be emphasised by taking into account a balanced bachelor of education curriculum in terms of science knowledge content and pedagogical skills. Induction of novice teachers into the teaching process needs to be strengthened and the mentor–mentee system between experienced teachers and new teachers needs to be implemented more openly. In addition, training programmes for graduate science teachers in higher learning institutions need to be strengthened as well. Furthermore, encouraging collaboration between universities and science teachers to improve science teacher training is also crucial.

The third recommendation is in terms of teaching, learning and assessment. Passive learning and teacher-centred teaching and learning should be minimised. This needs to be replaced with student-centred teaching and implementation of meaningful learning using various techniques and tools such as digital tools and cultural objects that can be found around students and schools. Furthermore, students' assessment should focus on improving conceptual understanding; science process skills; and creative, critical and innovative thinking skills rather than being totally examination-based.

Conclusion

For CRSP to be implemented in science lessons, teachers are required to have some knowledge of their students' lives outside of paper-and-pencil work and even outside of their classrooms. In order to engage students, teachers must adapt their teaching to the way in which those students learn rather than expect the students to adapt their learning to the way in which they are taught. Therefore, teachers need to know how to make ideas and knowledge meaningful for students and how to use students' funds of knowledge as tools to teach them. To sustain CRSP in the science classroom requires continuous effort and support for teachers' professional development and provision of autonomy to the teachers in terms of what and how to conduct the lesson in the classroom. Additionally, teachers need to have and adopt an innovative mindset as well for CRSP to be successfully implemented in the classroom.

References

Allen, B. A., & Boykin, A. W. (1992). African-American children and the educative process: Alleviating cultural discontinuity through prescriptive pedagogy. *School Psychology Review, 21*(4), 586–598.

Brown, C. L., & Bentley, M. (2004). ELLs: Children left behind in science class. *Academic Exchange Quarterly, 8*, 1–8.

Clark, A., Kjørholt, A., & Moss, P. (2005). *Beyond listening: Children's perspectives on early childhood services.* Policy Press.

Edwards, A. (2001). Researching pedagogy: A sociocultural agenda. *Pedagogy, Culture and Society, 9*(2), 161–186.

Fuchs, L. S., & Vaughn, S. (2012). Responsiveness-to-intervention: A decade later. *Journal of Learning Disabilities, 45*(3), 195–203. https://doi.org/10.1177/0022219412442150

Gainsburg, J. (2012). Why new mathematics teachers do or don't use practices emphasized in their credential program. *Journal of Mathematics Teacher Education, 15*, 1–21.

Gay, G. (2018). *Culturally responsive teaching: Theory, research, and practice* (3rd ed.). Teachers College Press.

Glynn, T., Cowie, B., Otrel-Cass, K., & Macfarlane, A. (2010). Culturally responsive pedagogy: Connecting New Zealand teachers of science with their Māori Students. *Australian Journal of Indigenous Education, 39*(1), 118–127.

González, N., Moll, L. C., Floyd-Tenery, M., Rivera, A., Rendon, P., Gonzales, R., & Amanti, C. (1993). *Teacher research on funds of knowledge: Learning from households.* National Center for Research on Cultural Diversity and Second Language Learning.

Gutiérrez, K., & Rogoff, B. (2003). Cultural ways of learning. *Educational Researcher, 35*(5), 19–25.

Kalantzis, M., & Cope, B. (2008). *New learning: Elements of a science of education.* Cambridge University Press.

Kontos, S., & Wilcox-Herzog, A. (1997). Teachers' interactions with children: Why are they so important? *Young Children, 52*(2), 4–12.

Ladson-Billings, G. (1994). What we can learn from multicultural education research. *Educational Leadership, 51*(8), 22–26.

Ladson-Billings, G. (2014). Culturally relevant pedagogy 2.0: A.k.a. the remix. *Harvard Educational Review, 84*, 74–84.

Lloyd, M. E. R. (2013). Transfer of practices and conceptions of teaching and learning mathematics. *Action in Teacher Education, 35*, 103–124.

Maitra, D. (2017). Funds of knowledge: An underrated tool for school literacy and student engagement. *International Journal of Society, Culture & Language, 5*(1), 94–102.

Moll, L. C., Amanti, C., Neff, D., & Gonzalez, N. (1992). Funds of knowledge for teaching: Using a qualitative approach to connect homes and classrooms. *Theory into Practice, 31*(2), 132–141.

Perso, T. (2012). *Cultural responsiveness and school education: With particular focus on Australia's first peoples; A review & synthesis of the literature.* Menzies School of Health Research, Centre for Child Development and Education, Darwin Northern Territory.

Phang, F. A., Abu, M. S., Bilal Ali, M., & Salleh, S. (2014). Faktor penyumbang kepada kemerosotan penyertaan pelajar dalam aliran sains: Satu analisis sorotan tesis (Factors contributing to the decline of student participation in science streams: An analysis of thesis review). *Sains Humanika*, *2*(4). https://doi.org/10.11113/sh.v2n4.469

Prebble, T., Hargraves, H., Leach, L., Naidoo, K., Suddaby, G., & Zepke, N. (2005). *Impact of student support services and academic development programmes on student outcomes in undergraduate tertiary study: A synthesis of the research*. April 14, 2015, www.educationcounts.govt.nz/publications/tertiary_education/5519

Shulman, L. S. (1987). Knowledge and teaching: Foundations of the new reform. *Harvard Educational Review*, *57*(1), 1–22.

Sleeter, C. (2012). Confronting the marginalization of culturally responsive pedagogy. *Urban Education*, *47*(3), 562–584. https://doi.org/10.1177/0042085911431472

Van den Bossche, P., & Segers, M. (2013). Transfer of training: Adding insight through social network analysis. *Educational Research Review*, *8*, 37–47.

Vygotsky, L. S. (1978). *Mind in society: The development of higher psychological processes*. Harvard University Press.

8 Case Study in Japan

Reflection from the Period for Integrated Studies

Kiyoyuki Ohshika and Murni Ramli

Introduction: Period for Integrated Study in Japanese Education System

Japan's national curriculum standard is formulated based on the Courses of Study (*gakushuushidouyouryou*) enacted by the Ministry of Education, Culture, Sports, Science and Technology (MEXT). The Courses of Study is revised every ten years, and the latest Courses of Study was announced for elementary and junior high schools in March 2017 and a year after (March 2018) for high schools. The elementary schools were expected to implement the new curriculum in 2020, junior high school in 2021, and high school in 2022. High schools are currently implementing the 2008 revised Courses of Study (MEXT, 2017).

The Courses of Study mainly defines the framework of subjects at each stage of school, from the elementary to high school. It also covers the goals of the subjects, and the general educational content. The Courses of Study is a subordinate ordinance of the School Education Law Enforcement Regulations stipulating the time allocation per subject in one academic year. Based on these regulations, the school formulates the curriculum by including consideration of the community issues as well (Nakayasu, 2016).

Period for Integrated Study in one of the subjects was first established based on the 1999 revision of the Courses of Study (MEXT, 1998). According to the report of the Curriculum Council in 2007, prior to the revision of the Courses of Study, each school was asked to secure time to develop specific educational activities that strengthen the use of ingenuity (MEXT, 2008a, 2008b). This regulation was established based on the proposal on allocating time for smoother cross-cutting and comprehensive learning beyond subjects to develop qualified students who can respond independently to all issues and changes in the society. In other words, the PIS activities must be the ones close to the students' daily life and the issues of the society where the students live, such as the issue of environment (Nakayasu, 2016).

PIS is established in all schools from elementary to high school. Although the time allocation for PIS in a year has changed, it will be continuously implemented in the new curriculum guidelines.

DOI: 10.4324/9781003168706-10

The Objectives of PIS

The objectives and content of PIS are as follows:

> To enable students to think in their own way about life through cross-synthetic studies and inquiry studies, while fostering the qualities and abilities needed to find their own tasks, to learn and think on their own, to make proactive decisions and to solve problems better. At the same time, enable students to acquire the habits of studying and thinking, and cultivate their commitment to problem solving and inquiry activities in a proactive, creative and cooperative manner.
>
> (MEXT, 2008c)

Each school sets goals and contents of overall PIS based on the school's goals. It means prior to the first semester; each school must decide the theme and the contents of the PIS of each grade. The time allocation or credits of PIS is 70 hours for Elementary school grades 3–6, 50 hours for junior high school grade 1, 70 hours for grades 2 and 3, and for high school, it is 3 to 6 hours.

The school must set learning tasks that are appropriate for achieving the goals of the school. The learning tasks include, for example, international understanding, information, environment, welfare or health, cross-cutting and comprehensive tasks, tasks based on students' interests, tasks according to the characteristics of the school, tasks related to profession and one's future, and so on. It is a learning task that is appropriate to study exploratorily, and it is an educationally valuable task in which learning, and awareness lead the students to think about one's way of life.

Learning Process, Ability, and Attitude in the PIS

In the PIS activities, the qualities, abilities, and attitudes that children/students are expected to own have been set up as follows:

There are three domains to be nurtured: qualities, abilities, and attitudes. These are the ways of learning, learning about oneself, and relationships with others and society. This policy is in line with the OECD's critical competencies. Furthermore, the examples of these abilities that should be utilised in the real world follow the learning method. The student can set up the way of learning, set the task, gather the information, think, analyse, and express oneself, such as the ability to criticise information or critical thinking ability, presentation ability, and others.

Additionally, it also includes the individual abilities that are expected to be possessed by the student, that is, abilities on decision-making, planning and execution, self-understanding, and future design (planning, action, self-control, etc.). In relation to others and society, students should develop understanding about other people and be able to co-operate and participate in the society by having communication and teamwork ability.

Exploratory learning is the fundamental learning approach that is promoted in the PIS (MEXT, 2008c). Through this learning approach, students will be nurtured to perform an iterative cycle of inquiry process which consists of the following consecutive procedures:

1   Set up their challenges or topic by focusing on daily life or issues in the society
2   Through the inquiry steps, they collect the information related to the topic
3   Sort out and analysing the data collected
4   Make a summary and representation
5   Repeat the process

The design of abilities construction in science education in Japan is based on the three abilities that are constructed, that is, reading ability and problem-solving ability which are abstracted from the PISA, and the inquiry ability which are promoted through PIS as listed:

1   Process of Reading Ability (PISA): (a) Access to information: pick out the information, (b) Integration and interpretation and (c) Deliberation and evaluation
2   Process of Problem-Solving Ability (PISA): (a) Inquiry and comprehension, (b) Representation and formularisation, (c) Plan and execution, and (d) Observation and deliberation
3   Process of Inquiry (PISA): (a) Set the challenges: have the consciousness of challenges through experiential activities, (b) Gather the information: pick out and accumulate the necessary information, (c) Sort out and analyse: sorting out and analysing the information, and (d) Summary and representation: Conclude, and represent a realisation, discovery, and idea

PIS Outcomes and Challenges

There is an annual educational survey in Japan on the National Academic Ability and the Situation of Learning conducted by the Curriculum Research Center of the National Institute for Educational Policy Research. This survey is to measure the abilities of grade 6 elementary school students and grade 9 junior high school students. The survey focuses on checking the subject course's implementation, learning motivation, learning methods, learning environment, and various aspects of students' lives (Tables 8.1 and 8.2).

The majority of elementary schools selected the theme "Problem according to the characteristics of the area and school" (98.7%) for each grade. And under that theme, the schools mostly chose the topic of "Environment" (89.9%) and "Local people's life" (89.5%). Meanwhile, junior high schools preferred the theme "Future occupation and self" (96.3%), where the schools developed the PIS project topics were "Career" (94.9%).

Table 8.1 Common learning themes in PIS selected by elementary and junior high schools

Themes	The theme selected by each grade (%)								
	3rd	4th	5th	6th	Total	7th	8th	9th	Total
A. Cross-cutting and comprehensive issues	71.7	79.1	81.2	81.1	88.3	55.1	53.5	56.9	63.2
B. Issues based on children's interests	57.0	59.4	60.6	67.1	74.4	42.5	44.2	48.8	55.5
C. Problem according to the characteristic of the area and school	93.3	82.5	82.1	78.3	98.7	74.2	60.6	57.8	79.9
D. Future of occupation and self	NA	NA	NA	NA	NA	71.1	92.2	86.4	96.3
E. Others	7.0	9.0	8.5	11.5	14.7	8.6	8.5	8.5	11.0

Source: Curriculum implementation/organisation status survey (NIER, 2013)

Table 8.2 Learning topics in PIS chosen by the elementary and junior high schools

Topics	Percentage of grades that selected the theme (%)								
	3rd	4th	5th	6th	Total	7th	8th	9th	Total
1. International understanding	37.8	37.6	28.6	43.6	65.8	16.1	17.6	25.6	33.1
2. Information	51.3	51.7	57.1	58.3	67.8	28.8	28.7	29.8	37.2
3. Environment	45.7	66.0	65.3	35.4	89.9	44.0	30.9	29.5	53.1
4. Welfare/Health	36.9	60.3	40.9	38.7	84.4	42.0	34.1	39.4	61.4
5. Local People's Life	80.6	53.1	52.5	47.2	89.5	NA	NA	NA	NA
6. Tradition and Culture	49.0	34.7	42.0	60.8	80.7	50.0	47.1	51.2	69.9
7. Disaster Prevention	11.2	16.9	14.3	14.3	26.5	23.1	21.9	22.1	27.9
8. Local development	NA	NA	NA	NA	NA	18.6	13.4	20.0	26.6
9. Career	NA	NA	NA	NA	NA	69.7	92.1	83.2	94.9
10. Others	15.2	22.3	24.8	35.3	42.7	16.8	17.9	18.4	22.7

Source: Curriculum implementation/organisation status survey (NIER, 2013)

In addition, the theme "welfare and health", "environment", "tradition and culture" are three topics that are also popular among the schools, both elementary and junior high schools.

Examples of PIS Initiatives

Cases of elementary school and junior high school as specific examples of efforts for comprehensive study time are introduced in the following section.

Elementary School

Ten years ago, Chiryu South Elementary School in Aichi Prefecture introduced learning activities to raise and cultivate creatures throughout the school. To systematically enhance learning about dealing with living things throughout the school and across subjects, the school has taken up and practiced "living environment learning" as a theme in the implementation of PIS for the entire school. Through this "living environment", the school aims to develop children's abilities from three perspectives (Figure 8.1).

At Chiryu South Elementary School, the environment for living things has been strongly introduced for nearly ten years. Elementary school children have been engaged in activities to develop their abilities and attitudes regarding living things through breeding and cultivation of living things and disseminating the results of those activities to the local people. In addition, by announcing their activities to junior students, they are also motivated to carry out activities for

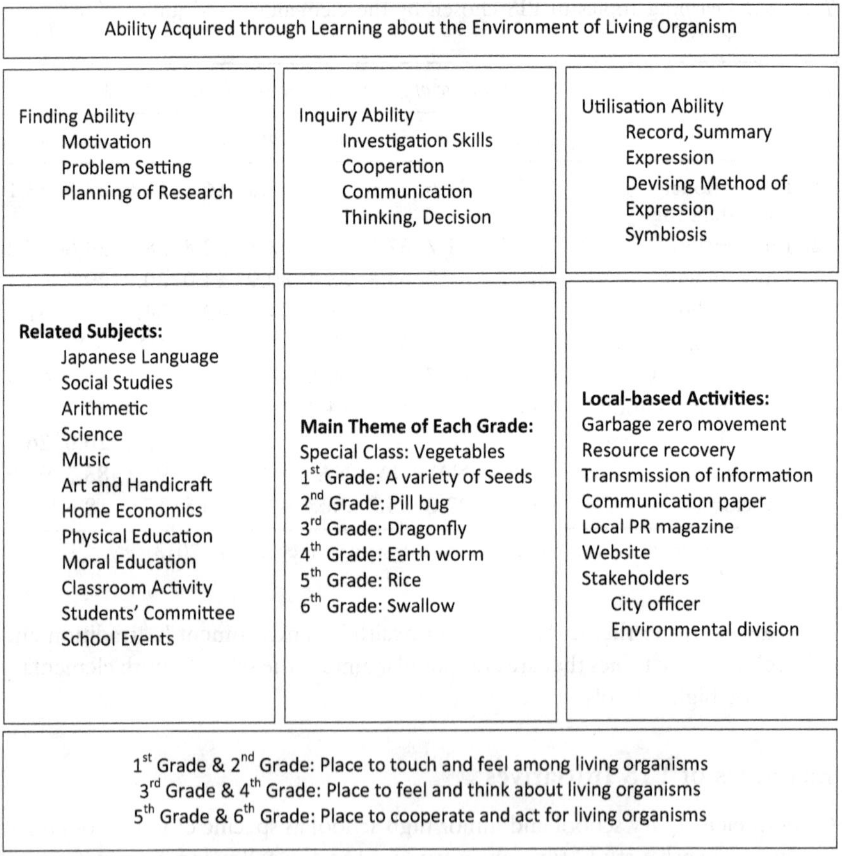

Ability Acquired through Learning about the Environment of Living Organism		
Finding Ability Motivation Problem Setting Planning of Research	**Inquiry Ability** Investigation Skills Cooperation Communication Thinking, Decision	Utilisation Ability Record, Summary Expression Devising Method of Expression Symbiosis
Related Subjects: Japanese Language Social Studies Arithmetic Science Music Art and Handicraft Home Economics Physical Education Moral Education Classroom Activity Students' Committee School Events	**Main Theme of Each Grade:** Special Class: Vegetables 1st Grade: A variety of Seeds 2nd Grade: Pill bug 3rd Grade: Dragonfly 4th Grade: Earth worm 5th Grade: Rice 6th Grade: Swallow	**Local-based Activities:** Garbage zero movement Resource recovery Transmission of information Communication paper Local PR magazine Website Stakeholders City officer Environmental division

1st Grade & 2nd Grade: Place to touch and feel among living organisms
3rd Grade & 4th Grade: Place to feel and think about living organisms
5th Grade & 6th Grade: Place to cooperate and act for living organisms

Figure 8.1 The curriculum of PIS at Chiryu South Elementary School

the following year. Moreover, by continuing this activity for six years, the school has deepened the understanding of the area by introducing the students to living things.

As a concrete example of PIS activities for grade 4 was learning about earthworms. Earthworms are known as soil organisms and live by ingesting and decomposing organic matter in the soil. They reduce the organic matter in the soil and convert it to rich soil. Through observing the breeding of earthworms, children learn what the earthworms eat, how they live, and how they change the soil. Meanwhile, the grade 1 students cultivate vegetables using the rich soil improved by the earthworms. Through this activity, the grade 4 students could be proud of what they have done as it is helpful for their juniors. Grader 4 students also learn about garbage in social studies. They learn composting as one of the treatments of kitchen waste and understand the decomposition mechanism by earthworms. It can be linked to the concept of recycling, in which garbage can be reused as a living resource and disposed of as garbage.

In Japanese elementary schools, lunch is served by the school. Elementary school students have many likes and dislikes of food and thus, many leftover foods ensue. These are treated like garbage, but at the same time, as the students learn about the recycling of leftover food by earthworms, it is possible to teach the students not to leave the food to waste as they can be recycled. This point can be included as "food education". Learning about earthworms is useful for various activities such as science, society, and eating habits.

In grade 6, learning advances under the theme of "swallows". In Japan, when the school season starts around April, swallows build nests around houses. Children search for swallow's nests in their area while investigating swallow's nests near the school. Through this investigation, the children will be able to communicate with the people who live near the school, where students will receive support from adults, and at the same time, they will be able to communicate their learning contents to adults. Swallows are migratory birds and travel abroad in winter. By learning the living period of swallows, domestic and overseas environments can be learned, and children can learn the global environment through swallows. As the final activities of the project, students are invited to present their investigation or exploration to the community at the forum facilitated by the school.

From 2018, as a development of this living things environment learning, the biotope plan started as a voluntary activity for the children. The Biotope is the area where the biological environment is uniform or called as a habitat or living place for specific animals and plants. The plan is as shown in Table 8.3.

The biotope creation plan is an idea that emerged during the grade 6 class, and after discussing with the principal, the plan was started. The students suggested that they need to create a place where they can raise the living things safely in the school and learn about the life cycles of those creatures. A biotope is also a place where various creatures can be observed at the school. Subsequently, since creating a biotope is impossible to be carried out by the children themselves,

Table 8.3 Yearly plan of Biotope in Chiryu South Elementary School

Date	6th Grade	3rd Grade
December	Discussion about the project of making a biotope	
January	Designing a biotope	Planning to make the "Insect Paradise"
	Submitting the idea to the principal	Planning the paradise of dragonfly, beetle, and butterfly
	• Making detailed blueprints	Requesting 6th graders to make put the "Insect Paradise" in their Hataori Biotop
	• Meeting with the Chiryu Gardening company	
	• Planning the *Hataori Biotope*	
February	Preparing the materials	Maintenance of the Insect Paradise
	Building the biotope	Planting cabbage
		Replanting dead trees
		Remaking a sign board
	Starting to activate the Biotope (For Active Times)	For active times
	Explore what we know about the biotope that we built	Exploring the dragonfly paradise, beetle forest, and butterfly paradise
	Petition for *Hataori Biotope* Project	
	Asking the Head of the area	
	Asking the Mayor	
March	Lively time (general announcement)	Lively time (general announcement)
	Completion ceremony for all students	Completion ceremony for 2nd-grade students

the principal then explained it to various people such as parents and local people and asked for cooperation from these people. After being told that the grade 6 students were planning a biotope, the local people came to work together with the students and were willing to cooperate with the school.

Meanwhile, the grade 3 students had another project, that is, learning about insects by building an insect park at school. The results after the activity were announced to all the students of grade 6, and students of grade 3 presented their insect project to the students of grade 2. The junior students listened to the announcement, and this motivated them to think about the next year's activities.

Although Chiryu South Elementary School set the learning about living things as the topic of their PIS, the learning material on the creatures is also connected to the learning of subjects. Therefore, it is also possible to exchange with other grades, or it could also be the content of learning by other grades. The spillover effect can be seen. Furthermore, it is possible to inform the local people around the school area about the school's activities and the growth and achievements of the children. Placing this PIS at the centre of the curriculum allowed it to be linked to the school-wide policy.

Junior High School

Shinkayama Junior High School in Aichi Prefecture is a rural junior high school that is far away from the city centre. The school district has mountains and valleys and a rich natural environment with several rivers flowing. This environmental background allows students at this school to learn a lot from this natural environment while cultivating the spirit of valuing and appreciating nature (Aichi Board of Education, 2017).

For those reasons, this school decided to adopt environmental study as a respect for the blessing of the natural environment. The activity aims to raise students' motivation to take action by grasping environmental problems as their own, while feeling a sense of crisis and urgency about the future environment, which will be nurtured systematically for three years. In addition, as part of the implementation of the movement "love for our hometown", the students would be actively and voluntarily engaged in various community activities such as cleaning activities, and including the conservation of local tradition called "*Sasay-uri* (bamboo lily)".

The school's primary focus on the environmental study is being promoted with the main theme of cultivating students who look at, think about, and work on the environment-development, as a part of ESD (Education for Sustainable Development) with an emphasis on "involvement, "connection," and "expansion".

The characteristics of students being nurtured at this school are as follows:

- Students who have acquired a wealth of knowledge about the environment.
- Students who can think accurately about what to do to protect the environment.
- Students who look back on their lives and actively work to protect the environment.

To promote this learning, the school has created its own "ESD plan". The plan for grades 7, 8, and 9 is as follows:

1   Grade 7: Think about a symbiotic society with living organisms

   - *Solidarity*: Be aware of the relationship between yourself, where you live, nature, society, and the future.
   - *Highly Diverse*: Feel the changes in the environment from the standpoint of living organisms and realise the significance of ecosystems and biodiversity.
   - *Incessancy*: Realise that the earth cannot be maintained due to the background of regional environmental changes on a global scale.
   - *Non-discriminatory*: Realise the difficulty of reconciling with the necessity of a symbiotic society between living organisms and humans.
   - *Commitment*: Understand the responsibility of the relationship between humans' life and biodiversity or environmental changes.
   - *Action Union*: Encourage ambition by realising the necessity of working together.

2   Grades 8 and 9: Think about coexistence for creating a sustainable society

- *Solidarity*: Try to act with an awareness of connections (people, society, nature, future).
- *Highly Diverse*: Think from various standpoints (consumers, producers, residents, the junior high student who have been affected by an earthquake disaster).
- *Incessancy*: Realise the necessity of eco-activity because of the limited resources and energies.
- *Non-discriminatory*: Consider measures for creating a sustainable society from a concrete perspective, such as electricity supply.
- *Commitment*: Value ethical perspective that transcends the generation and perceives it as one's issues.
- *Action Union*: Understand the necessity of working together with the community and society and encourage the ambition to take action in the future.

In addition, the following framework was created to determine the abilities and qualities students would acquire:

1   Learn to identify and set the problem:

- Critical thinking ability.
- Future design and planning ability.
- Multifaceted and comprehensive thinking ability.

2   Learn to act:

- Communication ability
- Ability to understand self and others,
- Activity environment maintenance ability

3   Learn to reflect:

- Environmental and social design ability
- Attitude to respect relation
- Attitude to participate

According to the plan in Table 8.3, the following six priority activities were carried out. The results are highlighted as follow:

1   *Sasayuri conservation activities*

To support grade 1 activities, the project to learn about biodiversity which aimed for coexistence with the natural environment was initiated, where the curriculum was reviewed and the position of PIS was clarified. As a result, the students were able to view the protection of Sasayuri as their affairs as well as view the current situation from the perspectives of "coexistence" and "nature conservation" and build their roles and ways of thinking on the future. The students also had new encounters with local NPO groups.

2 *Implementation of the environmental study programme*

The comprehensive annual learning plan created a few years ago had been persistently used in the same framework. The annual plan provided a general idea of the learning content and measures, but it was difficult to incorporate new tasks and activities. Instead of developing an individual annual plan for the grades 2 and 3, the plan was compiled at grade 2. Thus, a new programme was created and put into practice. The curriculum has also been improved by adopting the SDG key points as the theme of PIS. The theme of grade 1 is "biodiversity", while the theme of grade 2 is "low carbon society", and the theme of grade 3 is "sustainable society".

3 *Diversification of information dissemination and result announcement*

The school provided a forum for students to present their Sasayuri conservation activities and what they have learnt during the PIS. grade 3 students gave a presentation at the "Wildlife Conservation Activity Presentation Forum". The presentation was related to the learning in the environmental study that they have experienced for three years. Grade 2 students presented about "realising a low-carbon society" which they studied in the PIS. The results of PIS were not only for one's own learning but also for the development of a student's communication skills.

4 *Development of educational activities utilising information and communications technology*

The use of tablet PCs was able to deepen dialogue and learning of the students.

5 *Exchange of activities with junior high schools in disaster areas*

Shinkayama Junior High School continues to interact with junior high schools in the areas affected by the Great East Japan Earthquake. In the activities, they mainly talked about "current reconstruction", "cultural differences", and "regional conditions". Since Japan has diverse natural environments and cultures depending on the region, interacting with other students from various schools in other regions will help them reassess themselves, reaffirm similarities and peculiarities, and recognise their school and their own merits.

6 *Efforts for disaster prevention*

Presently, Japan is hit by various natural disasters such as earthquakes, volcanoes, and typhoons. Although students usually touch on information about disaster prevention, they have not taken it as their own. Through exchange activities with schools in the disaster areas and learning about disasters, students work towards enhancing disaster prevention education to prepare for natural disasters.

At Shinkayama Junior High School, the conservation of native plants, disaster prevention, and renewal of environmental study programmes as the PIS programme are developed from school activities to the school-wide curriculum. In Japanese junior high school, the teacher is charged with one subject. Therefore, it is difficult to see the overall growth of students as a whole compilation of all subjects. By creating a PIS-based curriculum, both students and teachers can clarify the state of growth.

Conclusion

The authors have introduced the efforts of schools to implement PIS at the elementary and junior high school levels. However, PIS is not about teaching something but more about the issues and characteristics of the school. As a result, it was found helpful in helping children and students acquire the various qualities and abilities that they need to develop and assimilate. The new Japanese Courses of Study, revised in 2017, places the development of the qualities and abilities of students at the centre of education and uses the process of inquiry as a method. The PIS, implemented at elementary or junior high school, will be changed to comprehensive science and mathematics inquiry. However, the training of students' inquiry continues to be emphasised. Currently, there is a demand for the development of citizens for the construction of a sustainable society. In the Japanese educational system, raising the qualities and abilities of children and students and nurturing their inquiry abilities are addressed to develop citizens who can respond to various issues. In that sense, PIS can be expected to play a central role.

References

Aichi Board of Education. (2017). *Activity Casebook of UNESCO school in aichi prefecture, Vol. 4*. August 27, 2019, www.pref.aichi.jp/uploaded/attachment/232910.pdf [In Japanese].

MEXT (1998). *The course of study for primary schools* (revised in 1998) [in Japanese]. Tokyo: MOE. August 27, 2019, www.mext.go.jp/a_menu/shotou/cs/1319941.htm. Accessed August 20, 2020 [in Japanese].

MEXT (2008a). *Manuals for the 2008 course of study in primary schools* [in Japanese]. Tokyo: MEXT. August 27, 2019, www.mext.go.jp/a_menu/shotou/new-cs/youryou/syokaisetsu/ [In Japanese].

MEXT (2008b). *Manuals for the 2008 course of study in lower secondary schools* [in Japanese]. Tokyo: MEXT. August 27, 2019, www.mext.go.jp/a_menu/shotou/new-cs/youryou/chukaisetsu/ [In Japanese].

MEXT (2008c). *Course of study for elementary school: Section of the period for integrated studies*. August 27, 2019, www.mext.go.jp/component/english/__icsFiles/afieldfile/2011/03/17/1303755_012.pdf

MEXT (2017). *Course of study for elementary school*. August 28, 2019, www.mext.go.jp/component/english/__icsFiles/afieldfile/2020/02/27/20200227-mxt_kyoiku02-100002604_1.pdf

Nakayasu, Chie (2016). School curriculum in Japan. *The Curriculum Journal, 27*(1), 134–150. https://doi.org/10.1080/09585176.2016.1144518

National Institute for Educational Policy Research (NIER). (2013). *Results of survey of school curriculum organisation and implementation status at the public elementary and junior high school in 2013 (Heisei 25-nendo kōritsu ko chūgakkō ni okeru kyōiku katei no hensei jisshi jōkyō chōsa no kekka ni tsuite)*. Ministry of Education, Culture, Sports, Science and Technology Japan, National Institute for Educational Policy Research. https://www.nier.go.jp/13chousakekkahoukoku

Part III

Way Forward

9 Culturally Responsive Pedagogy, STEM, and Gender Equity

Tackling Key Issues and Advancing the Research Field

Muhammad Abd Hadi Bunyamin and Izzah Mardhiya Mohammad Isa

Introduction

Traditionally, issues of gender equity are popular among Western nations, probably because the West is seen as more aggressive than the East in championing the issues, especially among scholars in the United States (e.g. Cimpian et al., 2020; Kong et al., 2020). Many vital concepts of equity (and social justice) in education were founded by American scholars, such as the notions of culturally responsive pedagogy (Ladson-Billings, 1994) and funds of knowledge (Gonzalez et al., 2005). However, it does not mean that the Eastern nations have put no effort into ensuring gender equity in education. Recent progress has given rise to increased interest in discussing and tackling gender equity issues among Eastern nations (e.g. Nugraha et al., 2020; Pilotti, 2021). Yet, these efforts are still insufficient; thus, more attempts are needed in Eastern nations to bring the agenda of gender equity to the centre of intellectual discourse.

In this chapter, the use of the term *gender* refers to boys and girls. In the West, "other" genders may be recognised; however, these are not given clear recognition in Eastern nations because of the cultural and religious acceptance in the East. A certain number of Eastern nations are predominantly Muslim-majority countries (Malaysia, Indonesia, Bangladesh, Brunei, Saudi Arabia, Pakistan, etc.) and their common belief on gender refers to only either boys or girls.

Gender Equity, STEM Education, and Culturally Responsive Pedagogy

In this study, the focus was on issues of gender equity in the context of education, specifically science, technology, engineering, and mathematics (STEM). The reason for putting the prime focus on STEM is that the authors expect many scholars have focused on issues of gender equity in STEM, probably due to the prominence of international assessments especially the Program for International Student Assessment (PISA) and the Trends in International Mathematics and Science Study (TIMSS), of which many nations around the globe have been involved in.

DOI: 10.4324/9781003168706-12

The authors adopted the framework of culturally responsive pedagogy (CRP) so as to ignite gender equity issues from a cultural perspective. CRP is a framework that was promoted by Ladson-Billings (1994). The framework is for educators to teach science or STEM to culturally diverse students (ethnicity, languages, etc.). Teachers who teach science or STEM from the CRP lens would be able to increase the relevancy of science or STEM for diverse students by capitalising on the diversities of the students' cultures for their benefit, especially by making them feel appreciated and empowered.

In this chapter, the primary reference is to the notion of CRP. Nevertheless, the authors also include other similar notions, namely culturally responsive teaching (CRT) (Gay, 2010) and culturally relevant education (CRE) (Aronson & Laughter, 2020). The reason is that CRP, CRT, and CRE all give a common focus on appreciating, including, and capitalising on the diverse cultures of students into teaching and learning.

The authors focused on gender equity when describing CRP/CRT/CRE because traditionally, these notions have been primarily focused on racial equity rather than gender equity. One particular contribution of this writing is that one can expand on the notion of CRP/CRT/CRE for the gender equity research area and shape the discourse of gender equity using the literature on CRP/CRT/CRE, which is less common.

This chapter mostly discusses the intersections of CRP/CRT/CRE, STEM, and gender equity because the intersections of the three areas can provide valuable insights on how issues of gender equity should be better addressed, researched, and solved from cultural viewpoints. To the best of the authors' knowledge, the past and recent literature have placed little focus on discussing and elucidating the intersections of these three areas. The initial analysis showed that many studies on CRP/CRT/CRE were conducted in the context of science or STEM without giving explicit focus on gender equity (e.g. Cole, 2015; Wilder & Axelrod, 2019; Yerrick & Ridgeway, 2017; Ziegler & Lehner, 2018).

Three guiding questions to direct the literature review on intersections of CRP/CRT/CRE, STEM, and gender equity are as follows:

1   How do the studies inform the issues of gender participation in STEM education, in actual school settings?
2   How do the studies published regarding STEM gender equity have possible interconnections with other relevant equity issues?
3   How are the studies published regarding gender equity inclusive in terms of the research participants involved?

The first question is pivotal because gender participation in STEM is one particular issue that is deemed long-standing. In the West, boys are more likely to participate in school STEM than girls. The issue still has merit because it is essential for boys and girls to get equal opportunities to learn and participate in STEM.

The second question is important because the notion of intersectionality is now available in the literature (Crenshaw, 1991). Focusing on gender equity in STEM may need to be looked at from the point of related issues such as race or socio-economic aspect. This is not to make the review complex but to make it more comprehensive in terms of the scope and to make it more potent in terms of impact.

The third question is vital because gender equity in STEM traditionally involves students. Nonetheless, with the efforts in making the issue more connected to society, other parties that can influence the issue should also be observed. Making the issue relevant to the community is beneficial as CRP concerns cultural matters where the community could be one of the particular parties that shapes cultural diversity.

Context of Study

This review adopted the systematic literature review (SLR) method. The authors searched publications with intersections of gender equity, STEM, and CRP/CRT/CRE. Primary databases were utilised to obtain relevant publications, and the databases include: (1) Web of Science, (2) Scopus, (3) SpringerLink Journal, (4) Wiley Online Library, (5) Taylor & Francis Online, and (6) MyCite. The Web of Science and Scopus databases were primarily referred to because they are the most common databases used by many (Mongeon & Paul-Hus, 2016). However, Mongeon and Paul-Hus (2016) have cautioned that relying on just the two databases can create biases and is even detrimental for social science research areas. Besides, the majority of studies published in the Web of Science and Scopus are in English. Thus, they would give a massive advantage for Western scholars who use English as the primary language of writing. Hence, the inclusion of other databases is essential to provide a balanced perspective, including the local databases such as MyCite (Malaysian Citation Index).

During the article search, the main keywords used were: "gender", "culturally responsive pedagogy", and "STEM education". They were used to ensure the search process would have a clear focus and would not become messy. The search was limited to the period between 2010 and 2020 because we wanted to review the recent research on gender equity in STEM. Nevertheless, the publications before 2010 were used as the review background because the central concept, namely CRP was established in 1994. It is believed that the past decades had created the foundation for the following decade in continuing the research on gender equity in STEM.

The author also limited the search to publications in the school context. The articles in higher education and workforce settings were excluded. The focus was given to school-context publications because gender equity issues appear to be less advocated in that specific setting.

Books published regarding gender equity in STEM were included. However, the focus was on books that looked at gender equity in STEM in the school context while those that focused on higher education and career contexts were excluded.

Initially, a total of 96 articles were found from all the databases used. From the initial result, the authors excluded publications that did not study the issues of gender equity in the STEM education context. The authors only covered STEM because STEM has become the primary area of study in education, especially considering the release of new standards or curricula around the globe that have emphasised STEM such as the Next Generation Science Standard (NGSS) in the United States in 2013 and the new curriculum of Standard Curriculum of Secondary Schools of Science in 2017 in Malaysia. In addition, Indonesia also released its new curriculum of science in 2013. Hence, many nations in the East and West have reformed their curricula of science which indicate a solid commitment to improving STEM education.

By excluding the publications that do not include STEM and gender, the authors managed to obtain 15 relevant publications (see Table 9.1). They were all read thoroughly to identify specific issues that the scholars studied. Even though the number of publications included for review is quite small, it could still provide valuable insights on gender equity in STEM from CRP's point of view. Many of the publications were excluded from the review because they did

Table 9.1 Publications reviewed and its main features

No.	Author/Year	Title	Journal/Book Name	Type of Study
1	Aronson and Laughter (2020)	The theory and practice of culturally relevant education: Expanding the conversation to include gender and sexuality equity	*Gender and Education*	Review
2	Ash and Wiggan (2018)	Race, multiculturalism and the role of science in teaching diversity: Towards a critical post-modern science pedagogy	*Multicultural Education Review*	Review
3	Carlone et al. (2015)	Agency amidst formidable structures: How girls perform gender in science class	*Journal of Research in Science Teaching*	Empirical
4	Dancstep and Sindorf (2018)	Creating a Female-Responsive Design Framework for STEM exhibits	*Curator: The Museum Journal*	Empirical
5	Hamlin (2013)	"Yo soy indígena": Identifying and using traditional ecological knowledge (TEK) to make the teaching of science culturally responsive for Maya girls	*Cultural Studies of Science Education*	Review

No.	Author/Year	Title	Journal/Book Name	Type of Study
6	Jackson (2015)	Perspectives and insights from preservice teachers of color on developing culturally responsive pedagogy at predominantly white institutions	*Action in Teacher Education*	Empirical
7	King Miller (2015)	Effective teachers: Culturally relevant teaching from the voices of Afro-Caribbean immigrant females in STEM	*SAGE Open*	Empirical
8	Kotluk and Kocakaya (2018)	Culturally relevant/responsive education: What do teachers think in Turkey?	*Journal of Ethnic and Cultural Studies*	Empirical
9	Milner (2016)	A Black male teacher's culturally responsive practices	*The Journal of Negro Education*	Empirical
10	Moote et al. (2020)	Comparing students' engineering and science aspirations from age 10 to 16: Investigating the role of gender, ethnicity, cultural capital, and attitudinal factors	*Journal of Engineering Education*	Empirical
11	Pringle (2020)	Toward a pedagogy of cultural relevance	*Researching Practitioner Inquiry as Professional Development* (Book Chapter)	Empirical
12	Samsudin et al. (2017)	Physics achievement in STEM project-based learning: A gender study	*Asia Pacific Journal of Educators and Education*	Empirical
13	Scantlebury (2014)	Gender matters: Building on the past, recognizing the present, and looking toward the future	*Handbook of Research on Science Education* (Book Chapter)	Review
14	Wieselmann et al. (2020)	"I just do what the boys tell me": Exploring small group student interactions in an integrated STEM unit	*Journal of Research in Science Teaching*	Empirical
15	Young et al. (2019).	Culturally relevant STEM out-of-school time: A rationale to support gifted girls of color	*Roeper Review*	Review

not conduct research on gender equity even though they did use CRP as their key framework and science or STEM as their research context (e.g. Cole, 2015; Wilder & Axelrod, 2019; Yerrick & Ridgeway, 2017; Ziegler & Lehner, 2018).

STEM Gender Equity in Actual School Settings

In STEM education, girls, in general, are underrepresented (Skolnik, 2015; Nimmesgern, 2016); nevertheless, girls are overrepresented in biological/life sciences. On the other hand, boys are dominant in the physical sciences (Hil et al., 2010). Based on the analysis performed on the published articles, we found that most scholars studied issues related to girls in STEM learning (e.g. Carlone et al., 2015; Dancstep & Sindorf, 2018; King Miller, 2015; Young et al., 2019).

In their review, Aronson and Laughter (2020) mentioned that girls' achievement in science was typically low in standardised tests. During middle school education, girls' achievement and interest in science start to show problems which continue when they enter high school and college. Nevertheless, the studies reviewed by Aronson and Laughter (2020) showed that the problems faced by girls in STEM education could be tackled with the use of curriculum and teaching that take into account the interests and learning preferences of girls. Using a girl-friendly curriculum and teaching strategies can positively change the attitudes of girls towards science/STEM (Dancstep & Sindorf, 2018; Hamlin, 2013).

Nonetheless, Aronson and Laughter (2020) did not specifically suggest steps in making girls more participative in learning science/STEM. The explanations provided were rather general and less insightful. A possible factor is that their review was performed across disciplinary subjects and not just science/STEM. Their review could be useful in understanding the general scenario of gender equity research across disciplines. However, the lack of specificity on STEM rendered their review less powerful and provided little meaning within the context of STEM education.

A few specific studies that could answer the question of how CRP/CRE/CRT should be placed in the actual teaching of science or STEM in schools are however available. Two studies (Milner, 2016; Pringle, 2020) provided in-depth explanations on making science teaching equitable in terms of gender; nevertheless, Pringle's study provided more details than Milner's. Pringle (2020) wrote a chapter describing how three teachers conducted science teaching with the CRE/CRT/CRP lens to tackle gender equity issues in science/STEM. Pringle compiled cases of three women science teachers, Mayra, Jennifer, Sara. Each of the teachers taught at different schools. Pringle reported what those women teachers did in teaching students with the lens of cultural relevance. The research approach that the teachers used was teacher research, which is different from the typical one, academic research. The teacher research carried out by the women teachers was less theoretical but more practical as the teachers primarily acted to counter real issues regarding equity in science learning among their students. The ways the teachers reported their research were largely descriptive. The teachers shared their specific strategies in teaching with the perspective of cultural relevance. For instance, one teacher, Mayra, focused on improving the learning of four African American girls. Each student had problems in learning such as having less interest in doing science activities or having problems completing work assigned by the teacher. Mayra adopted specific teaching strategies to tackle the students' problems such as using strategic grouping and cooperative learning. Mayra collected qualitative and quantitative data including field notes and pre- and post-assessment scores. The interventions were successful as the students showed greater engagement when learning in the class and an increase in the post-assessment scores.

Pringle (2020), in the same study, reported about another teacher, Sara, who shared her efforts in tackling issues related to gender equity. Unlike Mayra, Sara focused on African American male students. Additionally, Sara's classroom pre-dominantly comprised African American students. She mentioned that the African American boys had the lowest level of engagement and achievement in science. They usually did not complete the work assigned to them and were usually disruptive in classroom learning. As a result, they obtained a low score for the science subject. Eight students, all of whom were African American boys, were involved in the teacher research. Sara came up with a performance chart that showed the students' grades, missing and late assignments, and a note column. Sara used small-group discussion approaches to provide the students space to work together. However, Sara explicitly mentioned that she focused on building positive relationships with the students because she believed that it was the right way to ensure they would follow her instruction. Sara showed herself as a teacher who cared for the students and understood them. She then assigned the students a project for them to complete. As a result of her teaching approach, Sara reported that the students showed increased engagement in learning science.

Pringle's (2020) writing showed that with specific strategies designed to tackle equity issues in actual science classrooms, underrepresented groups' achievement and interest in science could be enhanced. The effort however must be intentional. The science teachers had explicitly shown their commitment to ensuring all students can learn, especially those who are underrepresented. In Pringle's study, the underrepresented group was African American students, both girls and boys. Some of the teachers focused on girls while others on boys.

Pringle's study could be a good example of how gender equity issues could be encountered in real science or STEM classrooms. Nonetheless, Pringle's study was mostly descriptive and less critical, which could possibly be attributed to the teacher research approach taken which was mainly to inform about the teachers' efforts to address and solve gender equity issues in STEM or science. Besides the role as the researcher, Pringle's other roles in the study were unclear. Was he guiding the teachers in conducting the teacher research? Nonetheless, what was obvious, as indicated in the final section of the writing, is that Pringle provided the interpretation of what the teachers had done in their classrooms with their underrepresented students.

Connections of Research on Gender Equity with Racial Equity in STEM

Across publications reviewed, one key pattern emerged: the studies on gender equity in STEM were conducted together with racial equity. Recently, many scholars have gone beyond gender perspectives per se, by taking into account other demographic information, especially race or ethnicity. Examples include Pringle (2020), Ash and Wiggan (2018), Jackson (2015), King Miller (2015), and Young et al. (2019). These studies have indicated the influence of intersectionality.

Crenshaw (1991) was the first to advocate the theory of intersectionality. The theory is about the notion of intertwined social categories that affect individuals in multiple ways and is influenced by the power structure. For instance, gender could be linked to race to explain why women of colour in the United States are underrepresented in STEM. In many publications that we reviewed, scholars such as Aronson and Laughter (2020), Pringle (2020), Ash and Wiggan (2018), Jackson (2015), King Miller (2015), and Young et al. (2019) have brought the notions of gender and race/ethnicity together in their writing. The trend indicates that recent studies in gender equity have expressed an interest in adopting the theory of intersectionality.

Scantlebury (2014) argued that intersectionality could be a theoretical framework for scholars to understand how gender can play its roles in advancing science/STEM education. Pringle's (2020) study has shown how intersectionality could be translated practically in actual science classrooms even though Pringle (2020) did not explicitly mention the intersectionality theory in the chapter written.

For years to come, research on gender equity may become more intertwined with racial equity. However, one particular question that arises is to what extent will the gender factor be given an equal emphasis with race? This question is critical because many scholars who brought together gender and race in their studies, review, or research, appear to focus more on race than gender. Two examples are provided in the following section.

Male, Black In-Service Teacher

Milner (2016) studied a Black male teacher's practice of CRP. The teacher, Jackson, was studied in terms of his thinking and practice of CRP. Milner argued that studying only a teacher was sufficient because he wanted to provide in-depth perspectives on how the idea of CRP/CRT/CRE was put in place and practised in the actual science classroom. Milner used a case study approach to collect and analyse the teacher's thinking and practice of CRP/CRT/CRE. The participant was a science and math teacher in a middle school in the United States. The middle school comprised almost 60% of coloured (African American) students, while White students were about 32%. It meant that the school was a Black-majority school.

Though his study included a Black male teacher, Milner was less explicit in telling how gender affected his study. We expected that a study like this should reveal the percentage of male and female students, yet Milner did not state this. It was pretty ironic because Milner studied a male teacher who is also an African American. Milner only mentioned the gender of the teachers in the school, where 7 teachers were male while 20 were female, with a total of 27 teachers altogether. Milner's findings indicated that the data were collected with a generic perspective regarding students' culture. Since the school involved in Milner's study was an African-American majority school in terms of student number, the teaching strategies that the teacher used might be closely relevant to Black students only. However, the gender aspect was not made explicit because Milner did not reveal the percentage of students in terms of their gender. Arguably, Milner should explain why he only included a male teacher in his study and why the students' percentage in the school in terms of gender was not revealed. At large, the study is deemed to have given more focus on race rather than gender.

Female, Black-Majority Pre-service Teachers

Jackson (2015) studied pre-service teachers of colour in relation to their perspectives of CRP. The pre-service teachers studied were predominantly from White institutions. A total of 30 participants were involved in the study. Twenty-four of those participants were females while the other six were males. Out of the thirty participants, twenty-five were African-American. Thus, the majority of the participants were Black Americans. Questionnaires were distributed to the participants and later analysed. The primary finding was that many of the participants had a limited understanding of CRP. Based on the fact that the pre-service teachers were studying at White-majority institutions, the main factor that might have shaped the research finding was race rather than gender.

Jackson (2015) did not explicitly take the gender perspective. Instead, it could be argued that the primary perspective in his study was race based on his recommendations regarding centring race issues in teacher education programmes. Jackson's study was unclear in terms of how gender affected his research findings despite him collecting the pre-service teachers' gender information when gathering the data. Thus, Jackson's study is thought to have focused more on race issues than gender.

Parents as Research Participants

Scholars (Aronson & Laughter, 2020; King Miller, 2015; Kotluk & Kocakaya, 2018; Young et al., 2019) have mentioned the roles of parents in making education accessible to all students, but not all of them carried out research in the context of STEM. For instance, King Miller (2015) studied five Afro-Caribbean women from Panama who immigrated to the United States. The study was conducted in New York City. All of them have had a career in STEM and could speak English and Spanish. King Miller conducted the study using the qualitative case study approach. Data were collected using interviews, qualitative surveys, field notes and observations, and artefacts and documents. The research participants represented the voices of educators (Nubia, Dorcas, and Afia) or parents (Andrea and Afia). Afia had dual roles in the study. Other scholars (Aronson & Laughter, 2020; Kotluk & Kocakaya, 2018; Young et al., 2019), however, did not directly study parents' voices in their research though one of them, Young et al. (2019), had mentioned that parents play the role of socialising agents for Black girls in increasing their science achievements.

Scholars in the research area of funds of knowledge such as Hensley (2005) have studied the roles of parents in making curricula more relevant for their children (students). In Hensley's study, teachers, parents, and children were all involved as participants. However, Hensley did not conduct the study from the gender perspective in STEM.

Why is it good to include parents in research on gender equity in STEM? First, parents are the individuals who are primarily responsible for their children's development at home. Since the birth of their children, parents provide informal education at home which may also include STEM elements such as teaching the children how to cook effectively and safely using the concept of heat and using suitable tools for cooking, including metals and heat insulators. Thus, their children learn STEM informally through daily activities at home. Parents can therefore observe directly how their children develop knowledge and skills of STEM.

Second, parents may significantly influence their children in selecting the science/STEM stream instead of the art stream for school studies. Parents, especially in Eastern nations, have been known to be the persons who may even decide the education paths of their children. Eastern children tend to follow their parents' views in deciding which education path to choose. If parents favour STEM, they can strongly influence their children's decision to choose STEM and often, parents are happy if their children follow their views. Parents are likely to have gender perspectives when developing their daughters or sons because educating girls and boys may require different approaches. The inclusion of parents in research on gender equity in STEM would thus greatly advance the research area.

Third, parents may have gender stereotypes. Being parents means they may need to take certain perspectives on how to develop their daughters or sons. Parents may balance their perspectives on developing girls and boys with sufficient knowledge and minimise gender biases and gender stereotypes. However, some parents may think that girls are less capable of being engineers and think boys are more capable of being one. Thus, having that perspective may discourage daughters from taking physics or physical science, which is a requirement in choosing engineering academic programmes or courses. Parents' perspectives that may contain gender biases or stereotypes should be discovered through research.

Conclusion

The authors suggest for scholars and teachers take several appropriate actions or steps to address, improve, or solve the concerns stemming from the three issues discovered in this study. The suggestions are to advance the research field, specifically in making the research agenda of gender equity in STEM more reachable in schools, in emphasising the research on gender equity with considerations on racial equity and in investigating parents' roles and influence in the issues of STEM gender equity.

First, the issues of gender equity in STEM are still happening in schools. It is suggested that scholars work with school teachers to make STEM gender equity issues more accessible in schools. Scholars and teachers can collaborate in teacher inquiry or practitioner research and publish journal articles or books together. An academic researcher can go beyond interpreting teachers' work by being a research partner of a teacher for the inquiry process. Collaboration between teachers and academic researchers may be more productive and desirable because teachers can learn from the academic researchers how to conduct a systematic inquiry by benefitting from the academic researchers' expertise. Academic researchers are often seen as theoretical experts, but they may not wholly understand the day-to-day scenario of STEM teaching and learning in schools, particularly on the issues of STEM gender equity. Similarly, academic researchers can learn from teachers of their real encounters with issues of gender equity in the actual science or STEM classrooms. The collaboration will make the solutions for STEM gender equity issues more accessible and feasible for schools and not just a debate in the literature. Accordingly, the study by Pringle (2020) could be a useful guide on how teachers and academic researchers can work together.

Second, scholars must have a clear direction when doing a study on STEM gender equity. Simply collecting demographic information regarding gender but not analysing them from a gender perspective reveals less commitment from the science/STEM education research scholars in advancing the research area. This claim aligns well with the statement by Hussenius et al. (2013) who discovered that fewer than 5% of research in science education is conducted on gender equity. Furthermore, while researching gender equity together with racial equity, a balance should be reached because many of the available studies have the tendency to not truly provide a good focus from gender equity's point of view. A balanced elaboration should be provided between gender and race. Future studies may be having intersectionality trends, especially in terms of gender and race. They may be more exciting and comprehensive; however, the perspective on gender should be provided deeply and explicitly. Otherwise, future studies may be losing insight on gender equity issues as racial issues seem more capable of attracting attention among scholars and teachers than gender. A possible reason could be that racial issues can easily trigger sentiments across races, especially from the political perspective. More efforts should be made to ensure gender equity issues can achieve equal status with racial equity issues; this could probably be achieved by providing more empirical data on how gender affects students' science achievement across races.

Third, very few scholars have researched on how parents influence boys and girls to choose and learn STEM. Scholars may learn from King Miller's (2015) study how inclusion of parents can enrich the literature on gender equity in science/STEM. Nevertheless, King Miller did not specifically highlight the roles of parents in influencing boys' and girls' learning in STEM. Hence, more studies should be conducted on parents, with a central focus on how parents shape boys' and girls' STEM learning informally at home in a way that is relevant to schools' STEM education. Additionally, the questions of how parents' perspectives on educating their daughters and sons may contain gender biases or stereotypes and how the perspectives affect STEM education of their children should also be addressed. To date, the inclusion of parents as research participants in the area of STEM gender equity research is rare. Future studies should consider this inclusion to advance the research area to the next level, that is, the community members.

References

Aronson, B., & Laughter, J. (2020). The theory and practice of culturally relevant education: Expanding the conversation to include gender and sexuality equity. *Gender and Education, 32*(2), 262–279.

Ash, A., & Wiggan, G. (2018). Race, multiculturalisms and the role of science in teaching diversity: Towards a critical post-modern science pedagogy. *Multicultural Education Review, 10*(2), 94–120.

Carlone, H. B., Johnson, A., & Scott, C. M. (2015). Agency amidst formidable structures: How girls perform gender in science class. *Journal of Research in Science Teaching, 52*(4), 474–488.

Cimpian, J. R., Kim, T. H., & McDermott, Z. T. (2020). Understanding persistent gender gaps in STEM. *Science, 368*(6497), 1317–1319.

Cole, M. W. (2015). Response to Marie Paz Morales' "influence of culture and language sensitive physics on science attitude achievement". *Cultural Studies of Science Education, 10*(4), 985–990.

Crenshaw, K. (1991). Mapping the margins: Identity politics, intersectionality, and violence against women. *Stanford Law Review, 43*(6), 1241–1299.

Dancstep, T., & Sindorf, L. (2018). Creating a female-responsive design framework for STEM exhibits. *Curator: The Museum Journal, 61*(3), 469–484.

Gay, G. (2010). *Culturally responsive teaching: Theory, research, and practice* (2nd ed.). Teachers College Press.

Gonzalez, N., Moll, L., & Amanti, C. (2005). Introduction: Theorizing practices. In N. Gonzalez, L. Moll, & C. Amanti (Eds.), *Funds of knowledge: Theorizing practices in households, communities, and classrooms* (pp. 1–24). Routledge.

Hamlin, M. L. (2013). "Yo soy indígena": Identifying and using traditional ecological knowledge (TEK) to make the teaching of science culturally responsive for Maya girls. *Cultural Studies of Science Education, 8*(4), 759–776.

Hensley, M. (2005). Empowering parents of multicultural backgrounds. In N. Gonzalez, L. Moll, & C. Amanti (Eds.), *Funds of knowledge: Theorizing practices in households, communities, and classrooms* (pp. 143–151). Routledge.

Hill, C., Corbett, C., & St. Rose, A. (2010). *Why so few? Women in science, technology, engineering, and mathematics.* AAUW.

Hussenius, A., Scantlebury, K., Andersson, K., & Gullberg, A. (2013). Ignoring half the sky: A feminist critique of science education's knowledge society. In N.

Mansour & R. Wegerif (Eds.), *Science education for diversity in knowledge society* (pp. 301–315). Springer.

Jackson, T. O. (2015). Perspectives and insights from preservice teachers of color on developing culturally responsive pedagogy at predominantly white institutions. *Action in Teacher Education, 37*(3), 223–237.

King Miller, B. A. (2015). Effective teachers: Culturally relevant teaching from the voices of Afro-Caribbean immigrant females in STEM. *SAGE Open, 5*(3), 2158244015603427.

Kong, S., Carroll, K., Lundberg, D., Omura, P., & Lepe, B. (2020). Reducing gender bias in STEM. *MIT Science Policy Review, 1*, 55–63.

Kotluk, N., & Kocakaya, S. (2018). Culturally relevant/responsive education: What do teachers think in Turkey? *Journal of Ethnic and Cultural Studies, 5*(2), 98–117.

Ladson-Billings, G. (1994). *The dreamkeepers: Successful teachers of African American children.* Jossey-Bass Publishers.

Milner, H. R. (2016). A black male teacher's culturally responsive practices. *The Journal of Negro Education, 85*(4), 417–432.

Mongeon, P., & Paul-Hus, A. (2016). The journal coverage of web of science and scopus: A comparative analysis. *Scientometrics, 106*(1), 213–228.

Moote, J., Archer, L., DeWitt, J., & MacLeod, E. (2020). Comparing students' engineering and science aspirations from age 10 to 16: Investigating the role of gender, ethnicity, cultural capital, and attitudinal factors. *Journal of Engineering Education, 109*(1), 34–51.

Nimmesgern, H. (2016). Why are women underrepresented in STEM fields? *Chemistry: A European Journal, 22*(11), 3529–3530.

Nugraha, I., Suranto, T., Kadarohman, A., Widodo, A., & Darmawan, I. G. (2020). The relation between gender, reasons to participate in STEM-related subjects, programs and the university supports on first-year university student's satisfaction: A structural equation model. *Journal of Science Learning, 3*(2), 117–123.

Pilotti, M. A. (2021). What lies beneath sustainable education? Predicting and tackling gender differences in STEM academic success. *Sustainability, 13*(4), 1671.

Pringle, R. M. (2020). Toward a pedagogy of cultural relevance. In Rose M. Pringle (Ed.), *Researching practitioner inquiry as professional development* (87–116). Springer Nature Switzerland AG. https://doi.org/10.1007/978-3-030-59550-0_6

Samsudin, M. A., Md Zain, A. N., Jamali, S. M., & Ebrahim, N. A. (2017). Physics achievement in STEM project-based learning (PjBL): A gender study. *Asia Pacific Journal of Educators and Education, 32*, 21–28.

Scantlebury, K. (2014). Gender matters: Building on the past, recognizing the present, and looking toward the future. In N. G. Lederman & S. K. Abell (Eds.), *Handbook of research on science education* (pp. 187–203). Routledge.

Skolnik, J. (2015). Why are girls and women underrepresented in STEM, and what can be done about it? *Science & Education, 24*, 1301–1306.

Wieselmann, J. R., Dare, E. A., Ring-Whalen, E. A., & Roehrig, G. H. (2020). "I just do what the boys tell me": Exploring small group student interactions in an integrated STEM unit. *Journal of Research in Science Teaching, 57*(1), 112–144.

Wilder, P., & Axelrod, Y. (2019). Humanizing disciplinary literacy pedagogy for Dinka refugee children. *Cultural Studies of Science Education, 14*(4), 1071–1077.

Yerrick, R., & Ridgeway, M. (2017). Culturally responsive pedagogy, science literacy, and urban underrepresented science students. In M. Milton (Ed.), *Inclusive principles and practices in literacy education* (pp. 69–86). Emerald Publishing Limited. https://doi.org/10.1108/S1479-363620170000011007

Young, J. L., Young, J. R., & Ford, D. Y. (2019). Culturally relevant STEM out-of-school time: A rationale to support gifted girls of color. *Roeper Review, 41*(1), 8–19.

Ziegler, J. R., & Lehner, E. (2018). Knowledge systems and the colonial legacies in African science education. *Cultural Studies of Science Education, 13*(4), 1101–1108.

10 Way Forward

Culturally Responsive Science Pedagogy for Developing Countries in Asia

Nurazidawati Mohamad Arsad, Nurfaradilla Mohamad Nasri, and Siti Nur Diyana Mahmud

Introduction

Malaysia and Indonesia are acknowledged as developing Asian countries with a range of races, religions, and ethnic groups. Meanwhile, Japan has always been recognised as a developed and modern country with a high proportion of ethnic homogeneity. Although Japan has 19 distinct racial and ethnic groups, the Japanese government refers to all Japanese people as "Japanese" (Oguma, 2021). Malaysia's population is made up of three distinct races: Malays, Chinese, and Indians. Besides these three groups, there are the *Orang Asli* who are the indigenous people living in Peninsular Malaysia. They do not constitute a homogeneous group as at least 95 subgroups of Orang Asli exist. In East Malaysia, the population of Sabah is made up of 32 ethnic groups, whereas the population of Sarawak consists of 27 ethnic groups. Another Asian country that has cultural diversity is Indonesia. This country is well-known as the world's largest island country and for its myriad ethnicities, with 633 ethnic groups (Ananta et al., 2014). Every country with a diverse race and ethnicity has inherited its own uniqtie ethnic culture.

The cultural differences among people in Malaysia and Indonesia lead to the urgent needs to implement the cultural relevance for students' learning, particularly in science education that emphasizes on the scientific thinking. However, the culture of scientific thinking from the Western view concerns more on knowledge and involves intentional information seeking (Kuhn, 2011) through deliberate activities, including exploring, making observations, asking questions, testing hypotheses, conducting experiments, recognizing patterns, making inferences, and evaluating evidence (Jirout & Zimmerman, 2015; Morris et al., 2012). This culture of scientific thinking is commonly recognised as empiricism and evidence-based, and considered burdensome by the local students since they derive their initial knowledge from the perspective of their local ethnic culture and religious belief that is difficult to be proven by the scientific activities. Therefore, to encourage the scientific thinking culture among local students, local educators should utilise culturally responsive science pedagogy (CRSP) based on the student's daily-life experiences as thoroughly discussed in Chapter 5 of this book on funds of knowledge (FoK), and not necessarily on the students'

DOI: 10.4324/9781003168706-13

respective traditional cultures. Thus, implementing culturally responsive teaching (CRT) strategy that utilises and connects to the students' FoK (Gay, 2018) will contribute to their meaningful learning and conceptual understanding.

CRSP was initially used in the Western nations in order to understand the challenges confronting the cultural differences in the inter-racial context for migrant people (Halim, 2021; Marosi et al., 2021). Meanwhile, whereas the various ethnic groups of Malaysia and Indonesia historically included a significant proportion of minority ethnic communities alongside the native population, there were also descendants of immigrants during the West's colonisation in the nineteenth century who remained. Yet, as time passes, these cultural differences have become cultures that already exist and is familiar in the local plural society. After all, culture is not based on the ethnicity, and an individual's ethnicity identity remains unchanged. In a broader sense, culture can be implying to their region, residential area, society, economic status, and gender. In this cultural context, learning from the case study in Japan on how they have adopted CRSP into their curriculum is a valuable opportunity.

Japan is well-known around the world as an Asian country that has had tremendous success in encouraging children to achieve academic excellence. Behind this success, the schools in Japan emphasise the importance of children's cultural and lived experience. The key factor in how elementary schools in Japan indirectly adopt CRSP during the Period on Integrated Studies (PIS) includes the provision of a task and the active use of a learning environment based on the characteristics of the local community and school, such as people's lives and consideration of community issues (MEXT, 2008). This supplementary course in the curriculum creates a sense of belonging between children and their school, and develops sociocultural consciousness through the bonding between children, school, and local community. Therefore, to achieve the best practice for equity in science teaching, especially for developing countries, CRSP should be aligned seamlessly with the curriculum and incorporated in teaching, learning, and assessment. Hence, a way forward for successful CRSP was discussed based on the case studies in this book.

Science Curriculum in Supporting CRSP

In recent years, many countries have embarked on curriculum reforms in order to keep up with global trends and to rectify any shortcomings that have halted development among children. Both Malaysia and Indonesia are no exception to this reform, resulting in the development and formation of new curricula so that the elementary and secondary school children can achieve quality and equity in education (United Nations, 2015) regardless of cultures. Thus, one of the best ways to achieve equity is through CRSP. Hence, in order to support CRSP in education setting, integrative subjects should be introduced in the curriculum.

Based on the case study, the education in Malaysia has already introduced integrative subjects in the curriculum. In 1988, "Human and Nature" became Malaysia's first interdisciplinary course subject, combining science, history, geography, health, and civics. It was intended for elementary schoolchildren aged 10 to 12 years old. This subject was designed to provide an awareness of how humans interacted with their environment. It has appeared to be ideal for children to learn because it integrates multi-disciplines. Even though this topic piqued children's attention at that age since the content was relevant to them, teachers faced difficulties to master the subject's content because they lacked the knowledge in disciplines associated with "Human and Nature" themes. Then, in a new curriculum transformation and with regard to the importance of science and technology, this subject was replaced by the science subject with additional pedagogy, including problem-solving-based learning, constructivism approach, and the science, technology, engineering, and mathematics (STEM) approach. Besides that, in order to be internationally competitive in the twenty-first century, children must be holistically developed with sound knowledge, skills, and values grounded in a strong national identity (Barghi et al., 2017).

Meanwhile, the science curriculum in Indonesia is an application-oriented education that emphasises the development of thinking skills, learning abilities, curiosity, and responsible attitude towards the natural and social environment. The integrated science curriculum in Indonesia integrates all aspects, namely attitude, knowledge, and skills. At the elementary school level, science is proposed to be taught as an integrated subject with Pancasila and Citizenship Education, Indonesian Language, and Mathematics. In the junior high school, science is developed as an integrative science subject, and not as a scientific discipline. Meanwhile, at the senior high school, three elective subjects are offered: biology, physics, and chemistry. Little is written about the integrative science in the junior high school basic competency documents, and not all basic competencies show integration.

In Japan, even though science subjects are offered in the curriculum, there is also another subject that thoroughly complements CRSP called "Period on Integrated Studies" (PIS). Although CRSP was not explicitly discussed or described in the curriculum, the main components of CRSP were embedded in PIS, and they appeared in the lessons that were relevant to the children and the community in which they lived. Besides that, children are provided with the potential for self-determination and empowerment by allowing them to identify their own tasks, learn and think independently, make proactive decisions, and solve issues more effectively. Additionally, children are encouraged to respond autonomously to all issues and changes in the society where they live, benefitting their socio-political consciousness development.

To promote CRSP, both Malaysian and Indonesian curriculum developers should highlight the concept of CRSP and its components in science curricula that lack cultural relevance as discussed in detail in the preceding chapter of this book. Despite the possibility of adding new subjects to implement CRSP, which consequently would necessitate the addition of other subjects, the best approach is to emphasise CRSP as a pedagogy in the science curriculum syllabus and to allot additional time for science subjects when teachers and students can apply the lessons. Additionally, the content of the curriculum standard document should emphasise how the lesson is executed using the CRSP framework, as described in Chapter 4, in a way that is responsive to the learners' needs and experiences, and that contributes to the local community.

Teaching, Learning, and Assessment

In both Malaysian and Indonesian case studies, the participants were novice science teachers with less than five years of experience in teaching science. Although they admitted that they were familiar with CRSP approaches because they reported that they practised CRSP in their science teaching, the critical analysis of their understanding of CRSP indicated that they mainly used daily-life experiences to explain scientific phenomena, reflecting the failure to engage the students in deep science learning. An important point to note is that, despite lacking an in-depth understanding of CRSP, both countries' educational policies clearly suggest the importance of connecting the students' daily life experiences with science learning, and this effort should be systematically as well as structurally planned by the science teachers to support meaningful science learning experiences. However, as convincingly suggested by Shaha and Ellsworth (2013), any governmental policies or initiatives should be closely monitored and evaluated by the authorities to ensure the effectiveness of the strategies. Due to the lack of monitoring aspects as well as support from the educational bodies, a great majority of the participating teachers reported that they rarely planned to include CRSP in their daily lesson plans. Most of them claimed that they tended to spontaneously integrate students' daily-life experiences into science learning without any initial planning. This practice has ignored the students' cultural background, which has long been deemed by researchers as one of the most influential factors shaping students' learning and understanding of the world.

Through this research project, the teacher participants were first introduced to the idea of CRSP prior to starting the proposed intervention programme of CRSP. It was surprising that most of the teacher participants were not aware of how to use FoK in the classroom as this aspect was not made clear during their teacher training programme. They also reported that the continuous professional development programme failed to address this important educational approach, which they deemed crucial in providing deep and meaningful learning experiences. Guided by close monitoring and support from researchers via online social media platforms, the teacher participants employed the co-planned intervention programme, and reported significant changes, especially in terms of the students' engagement during science learning. Furthermore, through the experimental research design, it was evident that the students' science process skills improved and most importantly, they showed a great interest in science learning.

Ultimately, all of the participants agreed that integrating various types of knowledge funds, such as academic and personal background knowledge, and accumulated life experiences and skills and knowledge used to navigate daily social contexts, was beneficial in leveraging local traditions to support science learning. They added that this culturally sensitive and responsive way of science teaching has not only proved to be successful in recognising students' outside world but also intellectually empowered the students to unleash their full potential. Although the concept of CRSP is relatively new in the Asian region, findings from this study echoed the findings of many researchers who strongly claimed that CRSP was an effective educational approach to address the quality, equity, and accessibility of education, especially for underserved students. Nevertheless, they repeatedly reported that trying to actively engage the students in science learning rather than relying on teacher-directed instruction was challenging due to the overwhelming teachers' workload and the large number of students per class. Therefore, it is suggested that the use of CRSP among science teachers should be a serious commitment by the Ministry of Education by adopting appropriate educational policies and pragmatic educational strategies.

Although all of the teacher participants both in the Indonesian and Malaysia case studies accepted CRSP approaches as one of the keys to promote student-centred learning, a majority of them reported that the assessment was less emphasised throughout the research project, and to some extent, they felt that the assessment practices were being ignored in CRSP. Thus, they were rather sceptical about the execution of CRSP since the assessment aspects were not made clear to them. Most importantly, as the country's practice of standardised assessment still dominated the educational system, all of the participants were reluctant to employ inquiry approaches that took more time and commitment to finish up the syllabus. However, as assessment plays an important role in shaping students' learning approaches, the nature of assessment practices should be meticulously planned.

The idea of constructive alignment, as indicated in the CRSP framework in Chapter 4, relies on the children's culture as a foundation to support the students' ability to achieve the learning outcomes. The children's cultural experiences include their prior knowledge and FoK, which come from their cultural experience, such as their daily-life activities, family inner culture, and work experience. Besides that, teaching approaches involve contextual teaching and learning and culturally responsive pedagogy (CRP), which emphasises the socio-political consciousness involved around the students' lives, families, school, and communities. Lastly, assessment approaches also need to be seriously considered. Constructive alignment is an integrative design for teaching, and the alignment between children's cultural, teaching-learning activities and assessment tasks is emphasised. The elements in the constructive alignment are similar to those in the PIS. The central step in designing learning is to define the intended learning outcomes, including what the students are supposed to learn and how they will demonstrate that learning has taken place (Biggs & Tang, 2011). Therefore, constructive alignment reflects the more general paradigm shift from teacher-centred teaching to student-centred teaching, and the student is seen as an active constructor knowledge (Tran et al., 2010). By choosing the appropriate assessment methods and tasks that adopt the socio-political consciousness perspective and aligning assessment with the intended learning outcomes and the teaching-learning activities, instructors can effectively guide students' study practices and enhance deep, and meaning-oriented learning (Biggs & Tang, 2011; Boud & Falchikov, 2006).

In summary, constructively aligned teaching is essentially a criterion-referenced system where the central elements, which are the intended learning outcomes, students' culture, teaching-learning activities and assessment, are aligned, and there is consistency throughout these elements. While the importance of constructive alignment is generally recognised, research investigating the effectiveness of the principle is rare and mostly theoretical in nature (Chadwick, 2004).

This study implies that constructively aligned teaching, including activating teaching and assessment methods, can especially support students who adopt an unreflective approach to science learning if they are not actively supported and encouraged to take an active role. The demands of the learning environment, including teaching-learning activities, can guide students' science learning in the desired direction. In this context, science teachers should always pay attention to how to actively engage the students, and shift the focus to what the students do and should do in order to learn. Active engagement is key to a student's success; hence, it is important to pay attention to how to engage students (Bolden et al., 2019). Note that assessment is a powerful way to guide student learning, and the focus on designing this should be continued in order to best support student learning.

Conclusion

Adopting CRSP as one of the pedagogies in the science curriculum and enabling integration within the subject may help students recognise the relevance of science in their everyday lives. Besides that, it helps to build a sense of belonging between the students and their school, and this sense is important for students, especially in developing countries with diverse cultural backgrounds. To ensure effective CRSP, constructive alignment or the CRSP framework should be introduced in the science curriculum and utilised in teaching, learning, and assessment. To employ constructive alignment in science teaching, teachers must acquire interdisciplinary science skills and cultural competence, with the students' culture being the foundation of constructive alignment. Therefore, teachers need to be provided with trainings to develop the science resource kits that incorporate both CRSP and contextual teaching. The training provided also needs to adopt the socio-political consciousness perspective in CRSP. Through this training, the teacher will acquire the knowledge and skills that facilitate the development of children's scientific thinking and make them aware of their surroundings, in addition to improving their local community. Socio-political consciousness has various meanings. In this case, it suggests to begin with what they can do within their capacity. For example, teaching about the importance of obtaining better inputs from their nearby entities, such as schools, parents, and community. Lastly, the authors argue that the main benefits of implementing

CRSP for students' learning are not only for the marginalised group but also to address the unintentional discrimination among those who are not marginalised.

References

Ananta, A., Arifin, E. N., Hasbullah, M. S., Handayani, N. B., & Pramono, A. (2014). *A new classification of Indonesia's ethnic Groups* (based on the 2010 Population Census). ISEAS Working Paper. ISEAS.

Barghi, R., Zakaria, Z., Hamzah, A., & Hashim, N. H. (2017). Heritage education in the primary school standard curriculum of Malaysia. *Teaching and Teacher Education*, *61*, 124–131. https://doi.org/10.1016/j.tate.2016.10.012

Biggs, J., & Tang, C. (2011). *Teaching for quality learning at university* (4th ed.). Open University Press.

Bolden, E. C., Oestreich, T. M., Kenny, M. J., & Yuhnke, B. T. (2019). Location, location, location: A comparison of student experience in a lecture hall to a small classroom using similar techniques. *Active Learning in Higher Education*, *20*(2), 139–152.

Boud, D., & Falchikov, N. (2006). Aligning assessment with long-term learning. *Assessment & Evaluation in Higher Education*, *31*(4), 399–413.

Chadwick, S. M. (2004). Curriculum development in orthodontic specialist registrar training: Can orthodontics achieve constructive alignment? *Journal of Orthodontics*, *31*(3), 267–274.

Gay, G. (2018). *Culturally responsive teaching: Theory, research, and practice* (3rd ed.). Teachers College Press.

Halim, A. (2021). The Indonesian curriculum: Does it retain culturally responsive teaching? *Journal of English Language and Culture*, *11*(1), 1–10. https://doi.org/10.30813/jelc.v11i1.2399

Jirout, J., & Zimmerman, C. (2015). Development of science process skills in the early childhood years. In *Research in early childhood science education* (pp. 143-165). Springer.

Kuhn, D. (2011). What is scientific thinking and how does it develop? In U. Goswami (Ed.), *The Wiley-Blackwell handbook of childhood cognitive development* (pp. 497–523). Wiley-Blackwell.

Marosi, N., Avraamidou, L., & Galani, L. (2021). Culturally relevant pedagogies in science education as a response to global migration. *SN Social Sciences*, *1*(147), 1–20. https://doi.org/10.1007/s43545-021-00159-w

MEXT (2008). *Course of study for elementary school: Section of the period for integrated studies*. August 23, 2021, www.mext.go.jp/component/english/__icsFiles/afieldfile/2011/03/17/1303755_012.pdf

Morris, B. J., Croker, S., Masnick, A. M., & Zimmerman, C. (2012). The emergence of scientific reasoning. In H. Kloos, B. J. Morris, & J. L. Amaral (Eds.), *Current topics in children's learning and cognition* (pp. 61–82). InTech.

Oguma, E. (2021). Racial and ethnic identities in Japan. In Michael Weiner (Ed.), *Routledge handbook of race and ethnicity in Asia*. Routledge. https://doi.org/10.4324/9781351246705-22

Shaha, S. H., & Ellsworth, H. (2013). Predictors of success for professional development: Linking student achievement to school and educator successes through on-demand, online professional learning. *Journal of Instructional Psychology*, *40*(1), 19–26.

Tran, N. D., Nguyen, T. T., & Mguyen, M. T. N. (2010). The standard of quality for HEIs in Vietnam: A step in the right direction? *Quality Assurance in Education*, *19*(2), 130–140.

United Nations. (2015). *Transforming our world: The 2030 Agenda for sustainable development.* https://sustainabledevelopment.un.org/sdg4

Index

138   *Index*

culturally relevant science teaching
  (CRST) 44, 82
culturally relevant teaching 4
culturally responsive instruction 91–92
culturally responsive pedagogy 3–5, 8,
  39, 44–45, 47, 48, 56, 57, 60, 65,
  89–92, 115–117, 133
culturally responsive science pedagogy
  (CSRP) 6, 9–13, 22, 31, 38, 45–46,
  67–70, 90–91, 129–135
culturally responsive teacher 51
culturally responsive teaching 4–6, 45,
  89, 94, 95, 97, 116, 130
culturally sustaining/revitalizing
  pedagogy (CSRP) 5–6, 9
cultural objects/artefacts 65
cultural pluralism 6, 77–87
cultural relevance 7, 120, 121, 129, 131
cultural sustainability 5
cultural variations 9
culture: Chinese culture 25; cultural
  identities 4, 5, 50; dominant
  culture 9, 11–12; features of 8–9;
  homogenous culture 28; minority
  culture 9, 29, 37, 40, 51, 58, 59, 60,
  83, 130; multicultural 3, 7, 9, 11, 18,
  80–81; non-dominant culture 4, 9;
  practices 8; in science 3, 13; in science
  education 3
curriculum: centralised curriculum
  20, 21; in culturally responsive
  pedagogy 4; in culturally responsive
  science pedagogy 5; curriculum
  differentiation 38–39; design 48, 51,
  60, 66–67; dominant narratives in
  13; Eurocentrism in 45; girl-friendly
  curriculum strategies 120; in Indonesia
  25, 79–80, 84, 86–87, 130–131;
  informative-dense curriculum 33;
  innovative-responsive curriculum 33;
  in Japan 28, 102, 108, 110, 111,
  112, 130–131; K13 curriculum 25;
  in Malaysia 20, 21, 22, 33, 36–39,
  90–94, 96–99, 118, 130–131; national
  curriculum 21, 25, 28, 37, 93,
  102; one-size-fits-all curriculum 48;
  prescribed curriculum 84; relevance
  48; science curriculum 9, 12, 21, 25,
  27, 28, 57, 59–60, 70, 83, 86, 92–94,
  97, 98–99, 130–131, 134; secondary
  education curriculum 20
curriculum based on Education Unit
  Level 79
curriculum differentiation (CD) 38–39

daily life experiences 22, 23, 80, 90, 91,
  96, 102, 104, 129, 132, 133
democratic society 77
developing countries 31, 34, 36, 50,
  130, 134
different ability backgrounds 39
disparities 31
distance learning 24
diversity 4, 18, 27, 29, 38, 39, 48, 57,
  59, 64, 78, 81, 83, 89, 92, 117, 129
dominant culture 9, 11–12
Dual Language Programme (DLP) 20
Dutch education 23, 79

Eastern nations 115, 124
ecosystems 64, 82, 86, 109
educational equity 31–40
educational resources 31, 34–36, 40
educational streaming 27
education systems: Indonesia 23–25,
  77–87; Japan 25–29, 28, 79,
  102–112; Malaysia 18–23, 28, 33–34,
  89–100
empower 4, 5, 6, 8, 12, 38, 44, 48, 52,
  57, 84, 116, 131, 133
environment: Asian environments 44;
  collaborative/challenging learning
  environment 91; community
  environment 85; conducive learning
  environment 59, 62; discriminatory/
  demoralizing 9, 11; living
  environment learning 106–107,
  109–111; natural environment
  106–107, 109–111, 131; safe/
  nurturing 45; school environment
  22, 52, 85, 104, 130, 134; social
  environment 131; student 56, 60, 65,
  67, 83, 89, 90, 95, 102, 104
environmental issues 7, 10, 82, 86, 104
environmental/social design
  ability 110
equal access 32, 33
equality 32, 83
equitable education 19–20, 27, 29, 38,
  45, 65
equitable science education 33
equity: achieving in classroom 89,
  130, 133; educational equity 31–40;
  equitable education 19–20, 27, 29,
  38, 45, 65; gender equity 115–126;
  principle of 80, 82; racial equity 116,
  122, 125; social justice and 6–7,
  12–13; teacher perceptions about
  equality and 83

lifelong leaning 20, 25
linguistic 4, 5, 6, 7, 13, 38, 45, 48, 50,
   52, 59, 64, 66, 89
linguistic scaffolding 64
living environment learning 106–107,
   109–111, 130
local communities 8, 22, 39, 50, 78,
   130, 132, 134
local content 79, 85
local funds of knowledge 90
local identity 33
local language 51, 78, 83, 85
local learning content 80
local people 10, 19, 104, 106, 108
logical positivism 44

Malaysia 18–23, 26, 28, 31–39, 49,
   50–51, 59, 61, 63–65, 89–100, 118,
   129–133
marginalised 4, 9, 11, 12, 13, 29, 47,
   51, 60, 83, 91, 135
minority 9, 29, 37, 40, 51, 58, 59, 60,
   83, 130
Modern Western Science (MWS) 10
Muhammadiyah 78
multicultural background 7
multicultural context 18
multicultural education 3, 7, 9, 11, 78,
   80–81
multiculturalism 7, 80
multidisciplinary 86, 87
multi-ethnic country 18, 21
multifaceted/comprehensive thinking
   ability 110
Muslim-majority countries 115

Nahdhatul Ulama 78
national curriculum 21, 25, 28, 37,
   93, 102
National Education Philosophy 19, 93
national exam 19, 22, 37, 49, 84, 87
national identity 131
national language 21, 24, 29
natural environment 131
non-dominant culture 4, 9
non-western 49, 50, 92
norms 8

objectivity 3, 10–11
one-size-fits-all curriculum 48

parents 40, 48, 60, 61, 63–64, 67, 90,
   93, 108, 123–124, 134
participate attitude 110

passive learning 99
Pedagogical Content Knowledge (PCK)
   94, 99
pedagogical framework 12, 27, 37, 45,
   46, 51, 67, 78, 79, 89, 91, 99, 134
pedagogical scaffolding 91
pedagogy *see* critical pedagogy;
   culturally relevant pedagogy (CRP);
   culturally relevant science pedagogy
   (CRSP); culturally responsive
   pedagogy; culturally sustaining/
   revitalizing pedagogy (CSRP);
   liberation pedagogy
Period for Integrated Study (PIS)
   102–108, 110–112, 130, 131, 133
phenomenological event 65
pluralism 6, 7, 64, 77–87, 130
policies 5, 27, 33, 36, 48, 50, 80, 81,
   86, 87, 94, 103, 108, 132, 133
policymakers 6
Pondok Pesantren 23
practical work 9, 39, 44, 59, 86, 120
prescribed curriculum 84
primary education 19, 20, 26, 27, 35
prior experiences 4, 66
prior knowledge 22, 45, 47, 49, 66, 80,
   84, 91, 133
problem-solving ability 104
problem-solving-based learning 131
process of inquiry 104
professional development 24, 51, 66,
   97, 99, 132
Programme for International Student
   Assessment (PISA) 22, 27, 31, 34,
   36, 104, 115

quality 19–20, 27, 29, 31, 32, 33–34,
   36, 37, 40, 61, 63, 79, 83, 87, 94,
   98, 130, 133

race 20, 39, 60, 80, 117, 122–123,
   125, 129
racial 4, 6, 7, 19, 45, 56, 89, 129, 130
racial equity 116, 122, 125
reading ability 104
reasonableness 66, 67
reciprocal peer learning 97
regional specificities 81
relevant 90, 91, 93, 95
religion 8, 22, 24, 39, 61, 65, 70, 78,
   80, 81, 93, 115, 129
repetitive learning styles 21
resourcefulness 66
respect 66

For Product Safety Concerns and Information please contact our EU
representative GPSR@taylorandfrancis.com
Taylor & Francis Verlag GmbH, Kaufingerstraße 24, 80331 München, Germany

9 780367 768263